Prevention and Treatment of Medical Diseases in Vulnerable Populations

Guest Editors

Gaspare Palaia
Alice Bruscolini
Massimo Ralli

MDPI

Basel • Beijing • Wuhan • Barcelona • Belgrade • Novi Sad • Cluj • Manchester

Guest Editors
Gaspare Palaia
Oral and Maxillofacial Sciences
Sapienza University of Rome
Rome
Italy

Alice Bruscolini
Department of Sense Organs
Sapienza University of Rome
Rome
Italy

Massimo Ralli
Department of Sense Organs
Sapienza Univertisy of Rome
Rome
Italy

Editorial Office
MDPI AG
Grosspeteranlage 5
4052 Basel, Switzerland

This is a reprint of the Special Issue, published open access by the journal *Applied Sciences* (ISSN 2076-3417), freely accessible at: www.mdpi.com/journal/applsci/special_issues/Prevention_Treatment_Medical_Diseases_Vulnerable_Populations.

For citation purposes, cite each article independently as indicated on the article page online and using the guide below:

Lastname, A.A.; Lastname, B.B. Article Title. *Journal Name* **Year**, *Volume Number*, Page Range.

ISBN 978-3-7258-3490-7 (Hbk)
ISBN 978-3-7258-3489-1 (PDF)
https://doi.org/10.3390/books978-3-7258-3489-1

Contents

applied sciences

MDPI

Editorial

Special Issue on Prevention and Treatment of Medical Diseases in Vulnerable Populations

Alice Bruscolini [1,*], Massimo Ralli [1] and Gaspare Palaia [2]

[1] Department of Sense Organs, Sapienza University of Rome, 00189 Rome, Italy; massimo.ralli@uniroma1.it
[2] Department of Odontostomatological and Maxillofacial Sciences, Sapienza University of Rome, 00189 Rome, Italy; gaspare.palaia@uniroma1.it
* Correspondence: alice.bruscolini@uniroma1.it

Several specific groups within the world's population have shown that they are particularly vulnerable to quality health care problems, differential experiences in the health system or increased burden of ill health. Among others, these groups include the poor, the elderly, women, children, those living with mental or physical disabilities, racial and ethnic minorities, and migrants. The vulnerability that these groups experience can usually be attributed to one or a combination of the following factors: economic status, geographical location, health status, or ability to communicate. This Special Issue aimed to collect and present research on medical diseases occurring in different groups of vulnerable people to improve clinical decision making and care of medical disorders that affect this target population. Fourteen papers (nine research papers and five review papers) from various fields concerning medical diseases, including ocular, oral, otolaryngologic, gastrointestinal disease, muscular dystrophy, knee laxity, osteoporosis and fibromyalgia, are published in this Special Issue. In addition, the results of the program to screen vulnerable populations for general health care status (one original article) or diseases in the head neck district (three papers) within the PROTECT project (Patologie del distRettO TEsta-Collo nei migranTi. Dalla formazione degli operatori alla diagnosi precoce e presa in carico del paziente: Network Odontoiatrico • Oftalmologico • Otorinolaringoiatrico • Maxillo-Facciale), funded by the Asylum, Migration, and Integration Fund (FAMI) 2014–2020 of the Ministry of the Interior and co-financed by the European Union are are reported in this Special Issue. Palaia et al. [1] reported that dental pathologies (especially untreated carious lesions or malocclusions) in migrants are unfortunately widespread and often neglected, highlighting the importance of prevention and early intervention. Bruscolini et al. [2] reported that about 50% of migrants carried out their first ophthalmological visit during the screening PROTECT project and emphasize the crucial role of early diagnosis of preventable visual impairment disease. Ralli et al. [3] reported that unilateral hearing loss was the most common otolaryngological condition and that snoring was the most frequent symptom reported during the PROTECT project, underlining the importance of early care to improve the quality of life of this disadvantaged population. Armentano et al. [4] conducted an observational study in patients with neurofibromatosis type 2, emphasizing the role of ocular manifestations in the early diagnosis of this rare disease. Comberiati et al. [5] described the ocular motility alterations in patients with Ehlers–Danlos Syndrome, suggesting the usefulness of a complete orthoptic evaluation in this rare disease. Gutierrez et al. [6] explored adherence to physical exercise in patients with fibromyalgia, focusing on the relationship between perceived and real execution of walking. Pergolini et al. [7] described oral adverse events in thirteen patients treated with different anti-programmed drugs (PD-1) and analyzed a possible correlation between the toxic manifestations of mucosal and dermatological to better manage these innovative anti-tumor therapies. Ahmad et al. [8] developed an Arabic word-in-noise screening tool and described its ability to early diagnose hearing deficits, especially in vulnerable populations. Iacoella et al. [9] screened more than

check for updates

Citation: Bruscolini, A.; Ralli, M.; Palaia, G. Special Issue on Prevention and Treatment of Medical Diseases in Vulnerable Populations. *Appl. Sci.* **2023**, *13*, 8886. https://doi.org/10.3390/app13158886

Received: 23 July 2023
Accepted: 29 July 2023
Published: 2 August 2023

four hundred individuals in disadvantaged neighbourhoods of Rome and reported the general low health conditions of this setting, encouraging a more efficient screening strategy. Dittrich et al. [10] revised and discussed the classification and diagnostic approach of osteoporosis in children with chronic diseases to improve the management of this vulnerable population. Iannucci et al. [11] analyzed the features and the clinical approach to uveitic glaucoma in childhood and reviewed the treatments available in the pediatric population. Sforza et al. [12] investigated and summarized the impact of blenderized tube feeding in comparison with conventional artificial tube feeding on improving gastrointestinal symptoms in children and adults and stressed the need of further experimental evidence on this topic. Cavallina et al. [13] collected and summarized the current knowledge of motor function in pediatric patients with congenital muscular dystrophies and discussed the tools used to evaluate it, stressing the need for a more accurate protocol. Moriceau et al. [14] investigated the role of the hormonal cycle in knee laxity and the risk of anterior cruciate ligament injury and reported, based on the literature, that there is no correlation between the menstrual cycle and the risk of ACL injury despite the menstrual cycle, due to hormonal fluctuation, seems to affect the laxity of the knee.

Although submissions to this Special Issue have been closed, it is essential to continue to address global challenges, such as promoting levels of well-being, especially in fragile and vulnerable populations, through accessible screening programmes and tailored protocols.

Funding: Some research (1, 2, 3) received external funding from the Asylum, Migration, and Integration Fund (AMIF) of the European Union, according to the code PROG-2010.

Acknowledgments: Thanks to all the authors and peer reviewers for their valuable contributions to this Special Issue, 'Prevention and Treatment of Medical Diseases in Vulnerable Populations'. I would also like to express my gratitude to all the staff and people involved in this Special Issue.

Conflicts of Interest: The author declares no conflict of interest.

References

1. Palaia, G.; Bossù, M.; Pranno, N.; Ralli, M.; Bruscolini, A.; Capocci, M.; Gaimari, G.; Macali, F.; Corridore, D.; Cotticelli, L.; et al. Oral Pathologies in Migrants: The Experience of the "PROTECT" Project in 3023 Patients. *Appl. Sci.* **2022**, *12*, 12621. [CrossRef]
2. Bruscolini, A.; Visioli, G.; Marenco, M.; Cherubini, V.; Comberiati, A.M.; Palaia, G.; Ralli, M.; Ottolenghi, L.; Lambiase, A.; Polimeni, A. Eye Health Screening in Migrant Population: Primary Care Experience in Lazio (Italy) from the PROTECT Project. *Appl. Sci.* **2023**, *13*, 3618. [CrossRef]
3. Ralli, M.; Colizza, A.; Russo, F.Y.; Palaia, G.; Angeletti, D.; Bruscolini, A.; Marinelli, A.; Bossù, M.; Ottolenghi, L.; de Vincentiis, M.; et al. Otolaryngology Conditions and Diseases in Migrants: The Experience of the PROTECT Project. *Appl. Sci.* **2023**, *13*, 2104. [CrossRef]
4. Armentano, M.; Lucchino, L.; Alisi, L.; Chicca, A.V.; Di Martino, V.; Miraglia, E.; Iannetti, L.; Comberiati, A.M.; Giustini, S.; Lambiase, A.; et al. Ophthalmic Manifestation in Neurofibromatosis Type 2. *Appl. Sci.* **2023**, *13*, 6304. [CrossRef]
5. Comberiati, A.M.; Iannetti, L.; Migliorini, R.; Armentano, M.; Graziani, M.; Celli, L.; Zambrano, A.; Celli, M.; Gharbiya, M.; Lambiase, A. Ocular Motility Abnormalities in Ehlers-Danlos Syndrome: An Observational Study. *Appl. Sci.* **2023**, *13*, 5240. [CrossRef]
6. Gutiérrez, L.; Lavín-Pérez, A.M.; Catalá, P.; Écija, C.; Collado-Mateo, D.; Gil-Arias, A.; Peñacoba, C. Patient's Perception and Real Execution of Walking as Physical Exercise: Looking at Self-Efficacy as a Key Variable in Adherence in Patients with Fibromyalgia. *Appl. Sci.* **2023**, *13*, 1191. [CrossRef]
7. Pergolini, D.; Botticelli, A.; Fascetti, R.; Rocchetti, F.; Cirillo, A.; Tenore, G.; Palaia, G.; Polimeni, A.; Romeo, U. Oral Immune-Related Adverse Events Associated with PD-1 Inhibitor Treatment: A Case Series. *Appl. Sci.* **2022**, *12*, 12994. [CrossRef]
8. Alokaily, A.O.; Alqabbani, A.F.; Aleid, A.; Alhussaini, K. Toward Accessible Hearing Care: The Development of a Versatile Arabic Word-in-Noise Screening Tool: A Pilot Study. *Appl. Sci.* **2022**, *12*, 12459. [CrossRef]
9. Iacoella, C.; De-Giorgio, F.; Palaia, G.; Ferraioli, M.; Arcangeli, A.; Ralli, M. Evaluation of General Health Status of Persons Living in Socio-Economically Disadvantaged Neighborhoods in a Large European Metropolitan City. *Appl. Sci.* **2022**, *12*, 7428. [CrossRef]
10. Dittrich, A.T.M.; Janssen, E.J.M.; Geelen, J.; Bouman, K.; Ward, L.M.; Draaisma, J.M.T. Diagnosis, Follow-Up and Therapy for Secondary Osteoporosis in Vulnerable Children: A Narrative Review. *Appl. Sci.* **2023**, *13*, 4491. [CrossRef]
11. Iannucci, V.; Manni, P.; Mecarelli, G.; Giammaria, S.; Giovannetti, F.; Lambiase, A.; Bruscolini, A. Childhood Uveitic Glaucoma: Complex Management in a Fragile Population. *Appl. Sci.* **2023**, *13*, 2205. [CrossRef]

12. Sforza, E.; Limongelli, D.; Giorgio, V.; Margiotta, G.; Proli, F.; Kuczynska, E.M.; Leoni, C.; Rigante, D.; Contaldo, I.; Veredice, C.; et al. The Impact of Blenderized Tube Feeding on Gastrointestinal Symptoms, a Scoping Review. *Appl. Sci.* **2023**, *13*, 2173. [CrossRef]
13. Cavallina, I.; D'Alessandro, R.; Brusa, C.; Panero, E.; Rolle, E.; Rossi, F.; Mongini, T.; Ricci, F.S. Motor Outcome Measures in Pediatric Patients with Congenital Muscular Dystrophies: A Scoping Review. *Appl. Sci.* **2023**, *13*, 1204. [CrossRef]
14. Moriceau, J.; Fevre, A.; Domínguez-Balmaseda, D.; González-de-la-Flor, Á.; Simón-Areces, J.; García-Pérez-de-Sevilla, G. The Influence of the Menstrual Cycle and Oral Contraceptives on Knee Laxity or Anterior Cruciate Ligament Injury Risk: A Systematic Review. *Appl. Sci.* **2022**, *12*, 12627. [CrossRef]

applied sciences

MDPI

Article

Ophthalmic Manifestation in Neurofibromatosis Type 2

Marta Armentano [1,†], Luca Lucchino [1,†], Ludovico Alisi [1], Antonio Valerio Chicca [1], Valerio Di Martino [1], Emanuele Miraglia [2], Ludovico Iannetti [3], Anna Maria Comberiati [3], Sandra Giustini [2], Alessandro Lambiase [1,*] and Antonietta Moramarco [1]

[1] Department of Sense Organs, Sapienza University of Rome, Policlinico Umberto I, Viale del Policlinico 155, 00161 Rome, Italy; marta.armentano@uniroma1.it (M.A.); luca.lucchino@uniroma1.it (L.L.); ludovico.alisi@uniroma1.it (L.A.)

[2] Department of Dermatology and Venereology, Sapienza University of Rome, Policlinico Umberto I, 00161 Rome, Italy

[3] Ophthalmology Unit, Head and Neck Department, Policlinico Umberto I University Hospital, Sapienza University of Rome, 00161 Rome, Italy

* Correspondence: alessandro.lambiase@uniroma1.it

† These authors contributed equally to this work.

Abstract: Neurofibromatosis type 2 (NF2) is a genetically determined tumor-predisposing syndrome. Ocular manifestations include cataracts, epiretinal membranes, retinal hamartomas, optic disk gliomas, and optic nerve sheath meningiomas. Moreover, optic disk edema, optical atrophy, motility disorders, pupil and lid dysfunction, and neurotrophic keratitis can be observed as indirect signs. An observational study was conducted with the aim to collect clinical data and describe the most frequent NF2 ocular manifestations. Fourteen patients affected by NF2, according to the Manchester criteria, were enrolled. All patients underwent complete ophthalmologic and orthoptic evaluation and a spectral domain optical coherence tomography. Ocular manifestations were present in all patients. The slit lamp evaluation of the anterior segment highlighted cataracts in five patients, keratitis in two patients, corneal leukoma in two patients, and corneal pannus in one patient. Fundus oculi and OCT evaluation identified epiretinal membranes in four patients, vitreoretinal tufts in three patients, optic nerve edema in one patient, and retinal hamartoma in one patient. Moreover, the orthoptic evaluation identified different types of ocular motility disorders in seven patients. This is a descriptive study of a rare disease with poor previous literature. Clinical data are shown, emphasizing the role of NF2-specific ophthalmological and orthoptic findings to help establish an early diagnosis.

Keywords: neurofibromatosis type 2 (NF2); merlin protein; cataract; epiretinal membrane; rare diseases

check for updates

Citation: Armentano, M.; Lucchino, L.; Alisi, L.; Chicca, A.V.; Di Martino, V.; Miraglia, E.; Iannetti, L.; Comberiati, A.M.; Giustini, S.; Lambiase, A.; et al. Ophthalmic Manifestation in Neurofibromatosis Type 2. *Appl. Sci.* **2023**, *13*, 6304. https://doi.org/10.3390/app13106304

Academic Editors: Francesco Cappello and Alexandros A. Lavdas

Received: 19 March 2023
Revised: 8 May 2023
Accepted: 18 May 2023
Published: 22 May 2023

1. Introduction

In 1882, von Recklinghausen described and classified neurofibromatosis type 1, neurofibromatosis type 2 (NF2), and schwannomatosis, under the common term "Neurofibromatosis" [1]. NF2 was later identified as a separate clinical entity by Gardner [2]. Neurofibromatosis type 2 is a tumor-predisposition syndrome characterized by the development of multiple schwannomas and meningiomas. The incidence is about 1 in 25,000 with a penetrance of 95%. The disease prevalence is estimated at 1 in 60.000. Thanks to earlier diagnoses and treatment developments, the life expectancy of these patients has improved, and the incidence of complications has decreased [3]. NF2 is caused by the inactivating mutations of the NF2 gene, located at q12.2 of chromosome 22, that result in a deficiency of the Merlin protein, a cytoskeletal protein with tumor suppressor properties [4].

NF2 is characterized by the development of bilateral vestibular schwannomas (VS), which lead to hearing loss, tinnitus, and imbalance. Other typical tumor types are schwannomas involving cranial, spinal, and peripheral nerves, and multiple meningiomas, both intracranial and intraspinal. Moreover, low-grade central nervous system (CNS) malignancies, such as ependymomas and gliomas, and skin tumors can occur [5].

The Manchester criteria, introduced in 1992, are the best-known diagnostic criteria for NF2 [6]. According to these criteria, a patient with suspicion of NF2 must meet one of the following clinical features:

1. Bilateral vestibular schwannomas;
2. First-degree relative with NF2 and unilateral vestibular schwannomas;
3. First-degree relative with NF2 or unilateral vestibular schwannomas and two among meningioma, cataract, glioma, neurofibroma, schwannoma, cerebral calcification;
4. Multiple meningiomas and two among unilateral vestibular schwannomas, cataract, glioma, neurofibroma, schwannoma, and cerebral calcification.

These criteria have been revised many times over the years. Recently, updates have been made, to distinguish between major and minor criteria and consider genetic analysis [7]. Specifically, the diagnostic criteria for NF2 have been recently expanded from the original Manchester criteria to include LZTR1 mutation testing in individuals with unilateral VS and other schwannomas. Moreover, two further changes were introduced, including NF2 mutational testing and the insertion of an age limitation of 70 for the development of bilateral VS if no other NF2 features are present.

NF2 phenotype severity is difficult to predict, and clinical manifestations appear variable among patients. Halliday et al. introduced a genetic severity score to predict the clinical expression of NF2. This score allows the connection of different genetic characteristics to a specific clinical disease presentation. This system consents to the individuation of tissue mosaicism form, classic and severe disease [8]. Interestingly, the genetic severity subtype correlated to NF2-related eye disease. Specifically, mutations associated with severe systemic disease resulted in greater visual morbidity at an earlier age.

To date, ocular pathologic findings of NF2 have been reported in a few retrospective series. A wide spectrum of ocular manifestations has been recognized to be specifically linked to NF2. Different clinical manifestations can occur in different age groups. Congenital cataracts, optic disc anomalies, and retinal hamartomas are responsible for the failure of binocular vision development in childhood. In middle-aged patients, a frequent cause of blindness is optic nerve sheath meningioma. In adulthood, visual functions can be compromised by ERMs or compression of the optic nerve due to intracranial tumors with the elevation of intracranial pressure. Visual impairment can also develop due to indirect NF2 complications, such as third, fourth, or sixth cranial nerve palsy, optic atrophy, or neurotrophic keratitis secondary to facial palsy [9,10]. This study aims to offer new insights into ophthalmological and orthoptic findings in patients with NF2 in the context of the paucity of available literature. We believe this work will raise awareness of this rare disease possibly leading to earlier diagnosis and assisting clinicians in patient care.

2. Materials and Methods

This single-center, observational, cross-sectional study was conducted at the University of Rome 'Sapienza', Umberto I Hospital, Italy, from June 2020 to February 2022. The research followed the Tenets of the Declaration of Helsinki, and written informed consent was obtained from all subjects and from parents in case of minor age. We included 14 consecutive patients between 16 and 78 years of age (mean age: 52 ± 10.1 years) with a diagnosis of NF2 based on the Manchester criteria [11,12]. NF2 patients were enrolled at the Regional Reference Centre for Rare Neurocutaneous Diseases of the Umberto I Hospital. All patients underwent 3-Tesla magnetic resonance imaging (MRI) of the brain, orbits, and spinal cord with gadolinium to assess the presence of vestibular schwannomas, schwannomas, and intracranial or spinal meningiomas.

Each subject underwent a detailed ophthalmological examination, including full medical history, measurement of best corrected visual acuity (BCVA) with Snellen's optometric chart at 5 m, biomicroscopic examination of the anterior segment, Goldmann applanation tonometry, indirect fundus biomicroscopy and spectral domain optical coherence tomography (SD-OCT). SD-OCT scans were obtained with the Spectralis OCT (Spectralis Family Acquisition Module, V 5.1.6.0; Heidelberg Engineering, Heidelberg, Germany) with the Heidelberg Eye Explorer (V1.6.2.0), whose axial resolution was 3.5 μm and transverse resolution was approximately 15/20 μm, using both the raster scan protocol (20° × 15°, 19 lines of scan) and the radial scan protocol (20°, 6 lines of scan), centered on the fovea. The orthoptic evaluation included inspection of eyes and head, Irvine test, cover–uncover test and alternate cover test, evaluation of ocular motility, assessment of objective convergence, and red filter test for diplopia.

3. Results

Demographic and relevant clinical features are listed in Table 1. In our study, the mean age was 53.1 years (range, 16 to 78 years) and five patients were males (10 eyes).

Table 1. Demographic characteristics, BCVA, and main ocular findings. RE: right eye; LE: left eye.

Patient	Sex	Age	Age at Diagnosis	VA RE	VA LE	NF2-Variant	Main Ocular Findings
1	F	55	51	1.00	1.00	Gardner variant	Lagophthalmos (LE)
2	F	58	37	0.90	0.80	Gardner variant	Epiretinal membrane grade 0, ptosis (LE), and exotropia
3	F	68	55	1.00	0.90	Gardner variant	Ptosis (LE) and exotropia
4	F	48	46	1.00	1.00	Gardner variant	Microstrabismus
5	F	63	50	1.00	1.00	Gardner variant	Lagophthalmos and superficial punctate keratitis (RE)
6	F	37	30	0.90	0.90	Gardner variant	Cataract (RE), paramacular epiretinal membrane grade 2 and optic nerve edema (RE and LE), and exotropia
7	F	62	28	1.00	1.00	Gardner variant	-
8	F	78	46	0.63	0.32	Gardner variant	Cataract (RE and LE) and retinal hamartoma (LE)
9	M	52	46	No light perception	0.50	Gardner variant	Corneal pannus and esotropia (RE) and corneal leukoma (LE)
10	M	59	59	1.00	1.00	Gardner variant	Lagophthalmos and superficial punctate keratitis (LE)
11	M	30	24	1.00	1.00	Gardner variant	Epiretinal membrane grade 1 and pseudophakia (RE), corneal leukoma and cataract (LE), and Duane syndrome type 1
12	M	73	71	1.00	0.32	Gardner variant	Cataract (LE) and microstrabismus
13	F	44	44	1.00	1.00	Gardner variant	Lagophthalmos (RE) and pseudophakia (LE)
14	M	16	10	1.00	0.03	Wishart variant	Cataract (LE), epiretinal membrane grade 2 (RE and LE), and microstrabismus

Ophthalmological manifestations were present in almost all patients and ranged widely from subtle retinal alterations, identified only by SD-OCT, to severe ocular involvement present at birth. The direct examination of ocular adnexa showed lagophthalmos in four patients (four eyes), eyelid ptosis in two patients (two eyes), and Duane syndrome type 1 in one patient (Table 2). One patient underwent tarsorrhaphy to treat severe lagophthalmos. In all cases, lagophthalmos was secondary to facial nerve palsy after surgical asportation of VS. All eyes with lagophthalmos developed exposure keratopathy, with reported two patients (two eyes) showing superficial punctate keratitis, two patients (two eyes) presenting with corneal opacification. One patient (one eye) developed vascularized corneal opacity due to recurrent infections. (Table 2).

Table 2. Ocular finding after inspection and observation of the anterior segment.

Ocular Findings	Number of Patients	%	Number of Eyes	%
Lagophthalmos	4	28.6%	4	14.3%
Ptosis of eyelid	2	14.3%	2	7.1%
Duane syndrome type 1	1	7.1%		
Pseudophakia	2	14.3%	2	7.1%
Cataract	5	35.7%	6	21.4%
Superficial punctate keratitis	2	14.3%	2	7.1%
Leukoma	2	14.3%	2	7.1%
Corneal pannus	1	7.1%	1	3.6%

The slit lamp evaluation of the anterior segment highlighted cataracts in five patients (six eyes). Three of them had posterior subcapsular cataracts, and two showed a cortico-nuclear type. Two patients aged 30 and 44 years were already pseudophakic. The IOP values were within the normal reference values in all patients (10–21 mmHg).

Fundus examination revealed the presence of epiretinal membranes in four patients (six eyes). Specifically, one patient (one eye) presented a grade 0 ERM, one (one eye) had a grade 1 and the other two (four eyes) had grade 2 ERMs according to the Gass classification (Figure 1A). Three patients (three eyes) showed the presence of retinal tuft (Figure 1B) and one patient was diagnosed with unilateral retinal hamartoma. MRI scans revealed no significant structural abnormality compatible with optic pathway gliomas or optic nerve sheath meningiomas. In addition MRI showed no signal change in the cranial nerves; however, one patient presented with bilateral optic nerve edema (Figure 2) associated with intracranial hypertension secondary to multiple compressive intracranial tumors, in particular meningiomas (Table 3).

Table 3. Findings on fundus examination.

Ocular Findings	Number of Patients	%	Number of Eyes	%
Epiretinal membrane	4	28.6%	6	21.4%
Vitreo retinal Tuft	3	21.4%	3	10.7%
Retinal hamartoma	1	7.1%	1	3.6%
Optic nerve edema	1	7.1%	2	7.1%

The orthoptic evaluation highlighted strabismus in 50% of our sample (seven patients). In detail, three patients (43%) presented microstrabismus, three patients (43%) exotropia, and one patient (14%) esotropia (Table 4). Diplopia was not observed in any case, and according to the orthoptic examinations, the three cases of exotropia were comitant. Moreover, one patient (7%) presented bilateral co-contraction of medial and lateral rectus, resulting in enophthalmos, reduced eyelid rim, and pseudo-ptosis. This clinical presentation was compatible with the diagnosis of Duane syndrome type 1. Physiological nystagmus in extreme gaze positions was found in eight patients (57%), three of them were orthophoric, and the remaining patients had a manifest ocular deviation.

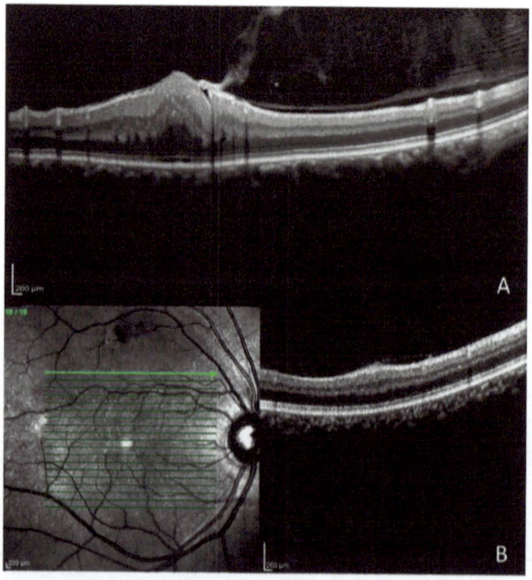

Figure 1. (**A**): Severe epiretinal membrane in a 16-year-old patient affected by the Wishart variant, OCT image. (**B**): Retinal tuft, OCT image.

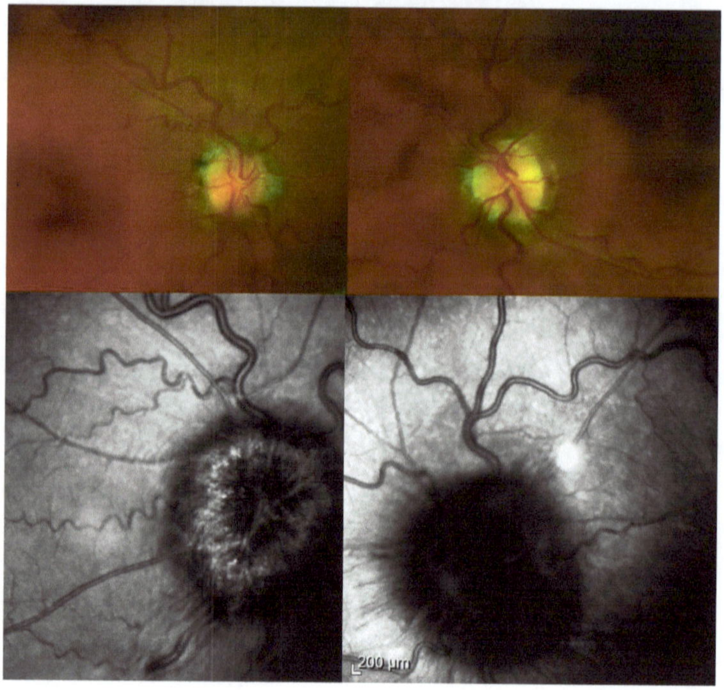

Figure 2. Optic disc edema associated with intracranial hypertension, fundus photography, and near-infrared image.

Table 4. Quantification of prism diopters in patients presenting strabismus. LCR: light corneal reflexes, CT: cover test, symm: symmetrical, PD: prism diopters.

Patient	LCR Near	LCR Distance	CT Near	CT Distance
2	−8°	−12°	−18 PD	−25 PD
3	−6°	−10°	−15 PD	−20 PD
4	symm	symm	+6 PD	+6 PD
6	−20°	−15°	−40 DP	−35 DP
9	+20°	+20°	+40 PD (Krimsky test)	+40 PD (Krimsky test)
12	symm	symm	+7 PD	+5 PD
14	symm	symm	+5 PD	+4 PD

4. Discussion

Type 2 neurofibromatosis is a complex, genetically determined pathology with a variable spectrum of manifestations. Genetical features have been characterized over time and brought together in a complex genetic severity score, which corresponds to clinical features. The tissue mosaic form (presumed and confirmed) includes patients with clinical signs of NF2 not confirmed by blood tests or patients with genetic alterations detectable only in specific tissues. The classic form includes mild and moderate NF2 clinical expression, with typical mutation identified in blood cells. The severe forms imply the presence of specific full truncating mutations involving exons 2–13, leading to aggressive phenotypes [8]. Age of diagnosis, age of onset, and the number of intracranial tumors, and mutation type remain the three major predictive factors of mortality in NF2 patients. Missense mutations have been linked to a lower risk of mortality. Moreover, clinical manifestations are more variable in patients with splice-site mutations [13]. In more than half of the cases, the most frequent symptoms at presentation are related to a vestibular schwannoma of the VIII cranial nerve. Spinal tumors are diagnosed in about 90% of patients, especially meningiomas in about 50% of patients [14].

The mild form of the disease (Gardner variant) occurs mainly with hearing loss (vestibular schwannoma) and presents a slower course and better prognosis than the severe early-onset form (Wishart variant). In the severe form, the ocular manifestations identified in childhood can lead to an early diagnosis of the disease as they often precede the other systemic manifestations. In the mild form, since the life expectancy of patients is 79 years (46 years in the severe form), the early identification of ocular manifestations is fundamental for preserving good visual function with targeted interventions [8].

In type 2 neurofibromatosis, the ocular involvement is frequent. In 2006, Bosch first classified the ocular features of the disease into NF2-specific and NF2-associated. In the NF2-specific group, he included posterior and cortical subcapsular/capsular cataracts (67% of patients), epiretinal membranes (40% of patients), retinal hamartomas (3% of patients), gliomas of the optic disc (13% of patients), and optic nerve meningiomas (27% of patients). In the NF2-associated group disc edema, optic atrophy, motility disorder, pupil dysfunction, lid dysfunction, reduced corneal sensation, exophthalmos, and exposure or neurotrophic keratopathy are included. The manifestations inserted in the latter group are caused by direct or indirect damage to the nervous structures determined by tumors [9].

Most recruited patients (13 out of 14) had late onset of symptoms (>20 years), corresponding to the mild form of the disease (Gardner variant). Among our cohort, the severe early onset form was identified in a single 16-year-old patient, who presented ERMs and subcapsular cataract, typically associated with the severe form.

In our study, in accordance with Bosch et al. the most frequently reported sign was cataract [9], which remains the only ophthalmological finding included among the NF2 diagnostic criteria [11].

Specifically, five of our patients (35.7%) presentedcataract, and two patients (14.3%) already underwent cataract surgery, amounting to seven patients (50%). Considering the young age of five of these seven patients, the development of cataract in their case can be plausibly attributed to NF2. We can underline that these patients presented a posterior subcapsular cataract, which is the form most frequently associated with the disease [15]. The other two patients with cataracts were, respectively 73 and 78 years old and the age of onset was not clear from the history, therefore it cannot be safely stated that the lens opacities in these patients are caused by NF2 or senile age.

Epiretinal membranes were the second most common sign in our sample after cataracts (28.6% of patients). ERMs in NF2 patients, at OCT evaluation, have atypical and distinctive characteristics, such as edges that project anteriorly into the vitreous despite an incomplete posterior vitreous detachment, a lack of cystoid macular edema, and an irregular and partially absent ILM [16]. These features distinguish it from idiopathic ERM or membranes associated with other diseases [16–18]. Histopathological studies highlighted that these abnormalities consist mainly of a mixture of Muller cells and astrocytes [19], supporting the hypothesis that ERMs represent hamartomas composed of Muller cells and justifying their morphological appearance [16]. OCT is the most sensitive method for studying this type of lesion and has allowed us to detect forms of various degrees: one of grade 0, one of grade 1, and four of grade 2 (macular pucker), according to the Gass classification [20]. Two young patients with macular pucker, aged 16 and 37, also showed subcapsular cataracts. The 16-year-old patient, as previously mentioned, was affected by a severe form of NF2. This observation agrees with several authors who showed a direct relationship between the prevalence of NF2-specific eye findings, the age of onset, and the severity of the genetic defect [8,9].

Another relevant finding detected in 21.4% of our patients were retinal tufts. This percentage appears to be inferior when compared with the recent literature. In a case–control study, Emmanouil et al. reported retinal tufts as the most common abnormalities of the central retina in 23 patients (43%), principally observed in patients with a severe form and not seen in controls, therefore showing a positive predictive value of 100% [21]. This difference can be explained because most of our patients have a mild form of the disease, as mentioned above.

Interestingly, Waisberg et al. reported the presence of choroidal abnormalities in NF2 patients. Choroidal abnormalities, visible as multiple, bright, patchy nodules in near-infrared reflectance (NIR) OCT, have recently been added to the revised diagnostic criteria for neurofibromatosis type 1 based on their high specificity and sensitivity [22]. In our sample, we have not found compatible lesions. Further studies on choroidal characterization in NF2 are needed [23,24].

As already mentioned, other ocular manifestations associated with NF2 are papilledema, optic atrophy, neurotrophic keratopathy, corneal keratitis and opacity, and ocular and eyelid motility disorders [9]. Among the non-specific ocular manifestations in our patients, we found changes affecting the cornea, eyelids, extrinsic muscles of the eye, and the optic nerve.

We highlighted eyelid ptosis in two patients (14.3%), lagophthalmos in four patients (28.6%), and strabismus in seven patients (50%). Eyelid ptosis is reported by Egan and other authors as a secondary manifestation of NF2, resulting from hypofunction of the oculomotor nerve [25]. Lagophthalmos is generally linked to post-surgical facial paralysis, following the removal of acoustic schwannomas. [9] This clinical feature was observed in our patients as a consequence of post-surgical VII nerve paralysis.

Corneal involvement was characterized by superficial punctate keratitis (two patients 14.3%), corneal leukoma (two patients 14.3%), and corneal pannus (one patient 7.1%). A post-surgical lagophthalmos, due to the removal of acoustic nerve neurinomas, was responsible for all these clinical conditions, as the incomplete closure of the eyelid rim determined corneal epithelial suffering.

One of our patients presented papilledema, due to increased intracranial pressure (7.1%). Bosch et al. described in their paper a 30% prevalence of optic disk edema, mostly caused by intracranial tumors or optic disk sheath meningiomas. The lowering of intracranial pressure, thanks to the surgical removal of cancerous masses, is not always sufficient to revert the progressive optic neuropathy. In these cases, patients can develop optic disk atrophy with irreversible loss of vision [9]. In the literature, papilledema is described in a small percentage of NF2 patients (<10%). The most frequent cerebral neoplasms associated with swelling of the optic disk are tumors in the posterior fossa, cranial nerve schwannomas, or large supratentorial meningiomas [26]. Our patient was diagnosed with multiple intra and extra-axial large and recurrent meningiomas, which required numerous surgical procedures.

Among our cohort, we also identified one case of retinal hamartoma (7.1%) in a patient with a moderate form of NF2. The lesion had a nasal peripapillary localization and was observable both during the fundoscopic examination and by OCT imaging. In the literature, retinal hamartomas have been reported with a variable frequency of 6–22% in NF2 patients [27]. These findings can be observed at the fundoscopic examination as whitish lesions, often involving the juxta papillary or macular area. OCT images are helpful to better characterize the extension and depth of such lesions. Retinal hamartomas are not specifically linked to NF2 but can be observed in other phakomatoses as well. Moreover, the prevalence of retinal hamartomas is higher in severe forms of NF2, often leading to a diagnosis based on ocular findings [28]. Parry et al. suggested that the presence of retinal hamartomas can be considered a clinical indicator of intermediate/severe forms of this pathology [5]. Furthermore, a frequent association between pigment epithelial and retinal hamartomas has been described [9,29].

Many of the reported manifestations can affect visual acuity. Within our sample, ERMs showed the greatest impact on BCVA, mainly when the fovea was involved, as we observed in one of our patients who had a BCVA of 0.03. Other causes of visual impairment in our NF2 patients were corneal opacities and cataracts.

Regarding ocular motility, in 2008, in a retrospective study, Feucht observed strabismus in more than half of patients with NF2, mostly determined by cranial nerve palsy (III-IV-VI) [30]. In the available literature, many studies identified ocular motility disorders in NF2 patients. Painter et al. retrospectively reviewed 83 patients affected by genetically characterized NF2. They diagnosed partial third nerve palsy in three children with severe mutation. Neuroimaging confirmed a third nerve schwannoma in all cases. Moreover, one patient developed a partial fourth nerve palsy [10]. Feucht et al. carried out a retrospective study of 73 NF2 patients. Different kinds of ocular motility disorders were identified in 52% of patients. About 22% of these patients presented cranial nerve palsy, due, in most cases, to intracranial tumors, and 26% showed concomitant strabismus, mainly horizontal [30]. Reggae et al. conducted a cross-sectional study on 49 NF2 patients, identifying oculomotor and abducens nerve palsy in six patients and concomitant strabismus in three patients [31]. Egan et al. identified monocular elevator paresis in four patients out of 29 affected by NF2, with dysfunction of both the superior rectus and inferior oblique muscles [25].

Barrett et al. described in a case report a two-year-old patient with recurrent third nerve palsy associated with schwannoma of the subarachnoid portion of the right third nerve [32]. Sokwala et al. described the case of a 21-year-old NF2 patient with a superior oblique muscle palsy associated with a sphenoid wing meningioma extended into the orbit [33].

The frequency of strabismus in our group was 50%, (three with microstrabismus, three with exotropia, and one with esotropia). The three cases of exotropia we described showed characteristics of comitant strabismus and were most likely attributable to a decompensated intermittent exotropia. Feucht et al. [30] reported comitant types associated with NF2 in 19 (50%) out of 38 cases of strabismus. The patient presenting with esotropia was affected by corneal pannus in the same eye with a visual acuity of no light perception. In this case, a plausible etiology could be a sensory deprivation.

5. Conclusions

This study describes ocular manifestations associated with NF2, confirming what was reported by other authors [19,30,31,34]. Our study showed cataracts, including posterior subcapsular cataracts and cortical wedge cataracts, as the most common ocular findings in NF2, followed by ERMs and retinal hamartomas, and emphasized the role of SD-OCT for the identification of subtle retinal lesions. Furthermore, the ophthalmological examination needs to be integrated with an orthoptic evaluation due to the high frequency of strabismus in NF2. The main limitation of our study is the small number of patients recruited, due to the rare occurrence of the disease. In the multidisciplinary team, the role of the ophthalmologist is strategic since it is well-reported that ophthalmological manifestations usually precede typical neurological symptoms in the pediatric population, thus making their exact recognition challenging for an early diagnosis and timely treatment.

Author Contributions: Conceptualization, A.L. and A.M.; methodology, A.V.C.; formal analysis, L.A.; investigation, M.A., L.L, V.D.M. and A.M.C.; data curation E.M. and S.G.; writing—original draft preparation, M.A. and L.L.; writing—review and editing, L.I., A.M., and A.L.; visualization, L.A.; supervision, E.M. and S.G.; project administration A.L. and A.M. All authors have read and agreed to the published version of the manuscript.

Funding: This research received no external funding.

Informed Consent Statement: Informed consent was obtained from all subjects involved in the study.

Data Availability Statement: The data that support the findings of this study are not publicly available to protect the privacy of research participants but are available from the corresponding author, A.L.

Conflicts of Interest: The authors declare no conflict of interest.

References

1. von Recklinghausen, F. *Ueber die Multiplen Fibrome der Haut und Ihre Beziehung zu den Multiplen Neuromen: Festschrift zur Feier des Fünfundzwanzigjährigen Bestehens des Pathologischen Instituts zu Berlin Herrn Rudolf Virchow*; A Hirschwald: Berlin, Germany, 1882.
2. Gardner, W.J.; Frazier, C.H. Bilateral acoustic neurofibromas: A clinical study and field survey of a family of five generations with bilateral deafness in thirty-eight members. *Arch. NeurPsych.* **1930**, *23*, 266–302. [CrossRef]
3. Evans, D.G.; Howard, E.; Giblin, C.; Clancy, T.; Spencer, H.; Huson, S.M.; Lalloo, F. Birth incidence and prevalence of tumor-prone syndromes: Estimates from a UK family genetic register service. *Am. J. Med. Genet. A* **2010**, *152A*, 327–332. [CrossRef] [PubMed]
4. Trofatter, J.A.; MacCollin, M.M.; Rutter, J.L.; Murrell, J.R.; Duyao, M.P.; Parry, D.M.; Eldridge, R.; Kley, N.; Menon, A.G.; Pulaski, K.; et al. A novel moesin-, ezrin-, radixin-like gene is a candidate for the neurofibromatosis 2 tumor suppressor. *Cell* **1993**, *72*, 791–800. [CrossRef] [PubMed]
5. Parry, D.M.; Eldridge, R.; Kaiser-Kupfer, M.I.; Bouzas, E.A.; Pikus, A.; Patronas, N. Neurofibromatosis 2 (NF2): Clinical characteristics of 63 affected individuals and clinical evidence for heterogeneity. *Am. J. Med. Genet.* **1994**, *52*, 450–461. [CrossRef]
6. Coy, S.; Rashid, R.; Stemmer-Rachamimov, A.; Santagata, S. An update on the CNS manifestations of neurofibromatosis type 2. *Acta Neuropathol.* **2020**, *139*, 643–665. [CrossRef]
7. Plotkin, S.R.; Messiaen, L.; Legius, E.; Pancza, P.; Avery, R.A.; Blakeley, J.O.; Babovic-Vuksanovic, D.; Ferner, R.; Fisher, M.J.; Friedman, J.M.; et al. Updated diagnostic criteria and nomenclature for neurofibromatosis type 2 and schwannomatosis: An international consensus recommendation. *Genet. Med. Off. J. Am. Coll. Med. Genet.* **2022**, *24*, 1967–1977. [CrossRef]
8. Halliday, D.; Emmanouil, B.; Pretorius, P.; Pancza, P.; Avery, R.A.; Blakeley, J.O.; Babovic-Vuksanovic, D.; Ferner, R.; Fisher, M.J.; Friedman, J.M.; et al. Genetic Severity Score predicts clinical phenotype in NF2. *J. Med. Genet.* **2017**, *54*, 657–664. [CrossRef]
9. Bosch, M.M.; Boltshauser, E.; Harpes, P.; Landau, K. Ophthalmologic findings and long-term course in patients with neurofibromatosis type 2. *Am. J. Ophthalmol.* **2006**, *141*, 1068–1077. [CrossRef]
10. Painter, S.L.; Sipkova, Z.; Emmanouil, B.; Halliday, D.; Parry, A.; Elston, J.S. Neurofibromatosis Type 2-Related Eye Disease Correlated With Genetic Severity Type. *J. Neuro-Ophthalmol. Off. J. N. Am. Neuro-Ophthalmol. Soc.* **2019**, *39*, 44–49. [CrossRef]
11. Smith, M.J.; Bowers, N.L.; Bulman, M.; Gokhale, C.; Wallace, A.J.; King, A.T.; Lloyd, S.K.; Rutherford, S.A.; Hammerbeck-Ward, C.L.; Freeman, S.R.; et al. Revisiting neurofibromatosis type 2 diagnostic criteria to exclude LZTR1-related schwannomatosis. *Neurology* **2017**, *88*, 87–92. [CrossRef]
12. Evans, D.G.; Huson, S.M.; Donnai, D.; Neary, W.; Blair, V.; Newton, V.; Strachan, T.; Harris, R. A genetic study of type 2 neurofibromatosis in the United Kingdom. II. Guidelines for genetic counselling. *J. Med. Genet.* **1992**, *29*, 847–852. [CrossRef] [PubMed]

13. Baser, M.E.; Friedman, J.M.; Aeschliman, D.; Joe, H.; Wallace, A.J.; Ramsden, R.T.; Evans, D.G.R. Predictors of the risk of mortality in neurofibromatosis 2. *Am. J. Hum. Genet.* **2002**, *71*, 715–723. [CrossRef] [PubMed]
14. Baser, M.E.; Kuramoto, L.; Joe, H.; Friedman, J.; Wallace, A.J.; Gillespie, J.E.; Ramsden, R.T.; Evans, G. Genotype-phenotype correlations for nervous system tumors in neurofibromatosis 2: A population-based study. *Am. J. Hum. Genet.* **2004**, *75*, 231–239. [CrossRef] [PubMed]
15. Bouzas, E.A.; Freidlin, V.; Parry, D.M.; Eldridge, R.; Kaiser-Kupfer, M.I. Lens opacities in neurofibromatosis 2: Further significant correlations. *Br. J. Ophthalmol.* **1993**, *77*, 354–357. [CrossRef]
16. Sisk, R.A.; Berrocal, A.M.; Schefler, A.C.; Dubovy, S.R.; Bauer, M.S. Epiretinal membranes indicate a severe phenotype of neurofibromatosis type 2. *Retina* **2010**, *30* (Suppl. S4), S51–S58. [CrossRef]
17. Waisberg, V.; Rodrigues, L.O.C.; Nehemy, M.B.; Bastos-Rodrigues, L.; de Miranda, D.M. Ocular alterations, molecular findings, and three novel pathological mutations in a series of NF2 patients. *Graefe's Arch. Clin. Exp. Ophthalmol. Albrecht. Graefes. Arch. Klin. Exp. Ophthalmol.* **2019**, *257*, 1453–1458. [CrossRef]
18. Schefler, A.C.; Dubovy, S.R.; Berrocal, A.M. Optical coherence tomography characteristics of epiretinal membranes in neurofibromatosis 2. *Ophthalmic. Surg. Lasers Imaging Off. J. Int. Soc. Imaging Eye* **2008**, *39*, 73–77. [CrossRef]
19. McLaughlin, M.E.; Pepin, S.M.; Maccollin, M.; Choopong, P.; Lessell, S. Ocular pathologic findings of neurofibromatosis type 2. *Arch. Ophthalmol.* **2007**, *125*, 389–394. [CrossRef]
20. Landau, K.; Yaşargil, G.M. Ocular fundus in neurofibromatosis type 2. *Br. J. Ophthalmol.* **1993**, *77*, 646–649. [CrossRef]
21. Emmanouil, B.; Wasik, M.; Charbel Issa, P.; Halliday, D.; Parry, A.; Sharma, S.M. Structural Abnormalities of the Central Retina in Neurofibromatosis Type 2. *Ophthalmic. Res.* **2022**, *65*, 77–85. [CrossRef]
22. Waisberg, V.; Rodrigues, L.O.C.; Nehemy, M.B.; Frasson, M.; de Miranda, D.M. Spectral-Domain Optical Coherence Tomography Findings in Neurofibromatosis Type 2. *Investig. Ophthalmol. Vis. Sci.* **2016**, *57*, 262–267. [CrossRef] [PubMed]
23. Legius, E.; Messiaen, L.; Wolkenstein, P.; Pancza, P.; Avery, R.A.; Berman, Y.; Blakeley, J.; Babovic-Vuksanovic, D.; Cunha, K.S.; Ferner, R.; et al. Revised diagnostic criteria for neurofibromatosis type 1 and Legius syndrome: An international consensus recommendation. *Genet. Med. Off. J. Am. Coll. Med. Genet.* **2021**, *23*, 1506–1513. [CrossRef] [PubMed]
24. Mallone, F.; Lucchino, L.; Giustini, S.; Lambiase, A.; Moramarco, A. An update on choroidal abnormalities and retinal microvascular changes in neurofibromatosis type 1. *Orphanet. J. Rare Dis.* **2022**, *17*, 223. [CrossRef] [PubMed]
25. Egan, R.A.; Thompson, C.R.; MacCollin, M.; Lessell, S. Monocular elevator paresis in neurofibromatosis type 2. *Neurology* **2001**, *56*, 1222–1224. [CrossRef]
26. Harada, T.; Sawamura, Y.; Ohashi, T.; Harada, C.; Shinmei, Y.; Yoshida, K.; Matsuda, H. Severe optic disc edema without hydrocephalus in neurofibromatosis 2. *Jpn. J. Ophthalmol.* **1998**, *42*, 381–384. [CrossRef]
27. Asthagiri, A.R.; Parry, D.M.; Butman, J.A.; Kim, H.J.; Tsilou, E.T.; Zhuang, Z.; Lonser, R.R. Neurofibromatosis type 2. *Lancet* **2009**, *373*, 1974–1986. [CrossRef]
28. Grant, E.A.; Trzupek, K.M.; Reiss, J.; Crow, K.; Messiaen, L.; Weleber, R.G. Combined retinal hamartomas leading to the diagnosis of neurofibromatosis type 2. *Ophthalmic. Genet.* **2008**, *29*, 133–138. [CrossRef]
29. Bouzas, E.A.; Parry, D.M.; Eldridge, R.; Kaiser-Kupfer, M.I. Familial occurrence of combined pigment epithelial and retinal hamartomas associated with neurofibromatosis 2. *Retina* **1992**, *12*, 103–107. [CrossRef]
30. Feucht, M.; Griffiths, B.; Niemüller, I.; Haase, W.; Richard, G.; Mautner, V.F. Neurofibromatosis 2 leads to higher incidence of strabismological and neuro-ophthalmological disorders. *Acta Ophthalmol.* **2008**, *86*, 882–886. [CrossRef]
31. Ragge, N.K.; Baser, M.E.; Klein, J.; Nechiporuk, A.; Sainz, J.; Pulst, S.-M.; Riccardi, V.M. Ocular abnormalities in neurofibromatosis 2. *Am. J. Ophthalmol.* **1995**, *120*, 634–641. [CrossRef]
32. Barrett, V.J.M.; Tan, M.H.; Elston, J.S. Recurrent third nerve palsy as the presenting feature of neurofibromatosis 2. *J. Neuro-Ophthalmol. Off. J. N. Am. Neuro-Ophthalmol. Soc.* **2012**, *32*, 329–331. [CrossRef] [PubMed]
33. Sokwala, A.; Knapp, C.; Gottlob, I. Neurofibromatosis type II presenting as vertical diplopia. *Strabismus* **2004**, *12*, 175–179. [CrossRef] [PubMed]
34. Kaye, L.D.; Rothner, A.D.; Beauchamp, G.R.; Meyers, S.M.; Estes, M.L. Ocular findings associated with neurofibromatosis type II. *Ophthalmology* **1992**, *99*, 1424–1429. [CrossRef] [PubMed]

applied sciences

MDPI

Article

Ocular Motility Abnormalities in Ehlers-Danlos Syndrome: An Observational Study

Anna Maria Comberiati [1], Ludovico Iannetti [1,*], Raffaele Migliorini [1], Marta Armentano [1], Marika Graziani [1], Luca Celli [2], Anna Zambrano [3], Mauro Celli [3], Magda Gharbiya [1] and Alessandro Lambiase [1]

1 Department of Sense Organs, University Sapienza, 00161 Rome, Italy
2 Section Endocrinology, Department of Internal Medicine, Amsterdam Bone Center, Amsterdam University Medical Centre, 1105 AZ Amsterdam, The Netherlands
3 Rare Bone Metabolism Center, AOU Policlinico Umberto I—Sapienza Università di Roma, 00161 Rome, Italy
* Correspondence: ludovicoiannetti@gmail.com

Abstract: Purpose: To evaluate ocular motility (OM) abnormalities associated with Ehlers-Danlos Syndrome (EDS). Materials and methods: In this cross-sectional observational study, patients with EDS underwent a complete orthoptic examination. The following orthoptic tests were performed: corneal light reflex test, stereoscopic test, cover test, OM assessment, evaluation of eye pain in different gaze positions and red filter test for diplopia. Results: The corneal light reflex test at 33 cm showed an intermittent divergent deviation in 31.7% of patients and an intermittent horizontal deviation associated with a vertical deviation in 4.9% of patients. A manifest strabismus was observed in 2.4% of patients, whereas 2.4% of patients showed a microstrabismus. The corneal light reflex test at 5 m revealed microstrabismus in 9.8% and manifest strabismus in 2.4% of our patients. Moreover, intermittent exotropia was observed in 2.4% of cases. No significant alterations involving the inferior rectus and the superior oblique muscles were observed. Significant associations were observed between medial rectus muscle deficit of both eyes with pain ($p = 0.020$) and diplopia ($p = 0.014$). Furthermore, a significant association between lateral rectus muscle alteration of both eyes and pain was observed ($p = 0.004$). Conclusions: Our results show various OM alterations in patients with EDS, specifically superior and medial rectus muscle hypofunction. A full orthoptic evaluation in these patients is recommendable to detect OM involvement and possible ligamentous laxity changes over time through an accurate OM assessment.

Keywords: Ehlers-Danlos Syndrome; collagen alteration; ocular motility; strabismus

check for updates

Citation: Comberiati, A.M.; Iannetti, L.; Migliorini, R.; Armentano, M.; Graziani, M.; Celli, L.; Zambrano, A.; Celli, M.; Gharbiya, M.; Lambiase, A. Ocular Motility Abnormalities in Ehlers-Danlos Syndrome: An Observational Study. *Appl. Sci.* **2023**, *13*, 5240. https://doi.org/10.3390/app13095240

Academic Editors: Laura Cercenelli, Alice Bruscolini, Massimo Ralli and Gaspare Palaia

Received: 4 March 2023
Revised: 13 April 2023
Accepted: 21 April 2023
Published: 22 April 2023

1. Introduction

Ehlers-Danlos Syndrome (EDS) is a rare disease that includes a clinically and genetically heterogeneous variety of connective tissue diseases. This hereditary pathology affects predominantly the skin, joints, and the musculoskeletal apparatus. Nowadays, 13 forms of EDS are recognized, and pathogenetic mutations have been identified in 20 different genes. The global incidence and prevalence of the disease is unclear. The incidence of one of the most frequent variants, the classic type (cEDS), has been estimated at around 1:20.00. The vascular one (vEDS) is one of the rarest forms and its incidence is approximated to be between 1:50.00 and 1:200.000. The most frequent mutations concern genes encoding for fibrillary collagen types I, III and IV, enzymes linked to the biosynthetic process of these proteins or other components of the extracellular matrix [1]. Many classifications have been proposed over time, the most famous being the Villefranche classification, introduced in 1997 [2]. The most recent revision of the EDS classification was introduced in 2017 and includes 13 types with well-characterized genetic features. Moreover, this new classification provides a detailed list of major and minor criteria to help the clinician during the diagnostic process [3]. Thanks to DNA sequencing analysis, for the majority of EDS variants, a genetic cause has been identified. cEDS is determined by heterozygous pathogenetic variants in COL5A1 or COL5A2, leading to haploinsufficiency in type V procollagen. vEDS is caused by defects in type III procollagen resulting from COL3A1 mutations. Different types of alterations involving type I procollagen are responsible for Arthrochalasia EDS, Dermatosparaxis EDS and Cardiac Valvular EDS. Defects of collagen crosslinking or folding are detectable in Kyphoscoliotic EDS forms. Classical-like EDS and Myopathic EDS are attributable to dysfunction in extracellular matrix (ECM) components. Musculocontractural EDS and Spondylodysplastic EDS are attributable to defects in glycosaminoglycans' biosynthetic process. In Periodontal EDS, an abnormal interaction between the complement and ECM has been recognized. The brittle cornea syndrome is attributable to the mutation of a zinc finger protein (ZNF469) with unknown function and a transcriptional regulator (PRDM5) leading to dysregulation in many genes, such as COL4A, COL11A1 and genes involved in ECM deposition. Regarding the hypermobile variant, a genetic cause has not been identified [1]. All EDS forms share some pathognomonic clinical characteristics, particularly joint hypermobility, cutaneous hyperlaxity, atrophic scarring, skin fragility and the tendency to develop spontaneous ecchymosis [3]. The classic and hypermobile variants are the most frequently diagnosed, representing around 90% of all forms. The vEDS is one of the rarest, being identified in less than 5% of cases. As a result of the clinical overlapping of EDS with other inheritable connective tissue disorders, a purely clinical diagnosis is often difficult to perform. For this reason, the possibility to confirm the diagnosis with genetic analysis is fundamental [4]. At the same time, for many patients with a clear EDS clinical presentation, a genetic anomaly is not detectable [5].

EDS ophthalmological manifestations involve corneal, conjunctival, orbital and vascular structures. Corneal abnormalities described in the literature are microcornea, megalocornea, corneal thickness and curvature alterations, surface irregularity and decreased corneal sensitivity. The progressive corneal thinning and steepening can lead to spontaneous ruptures with subsequent scarring and to elevated astigmatism and keratoconus [6,7]. These alterations are linked to an important reduction in type 5 collagen in the corneal stroma and reduced density of collagen fibrils. The finding of larger-diameter fibrils and irregular organization can also affect corneal transparency [8]. The corneal involvement is frequently encountered in cEDS patients, in the presence of type V collagen mutations. Conjunctival abnormalities observed in EDS are severe conjunctivochalasis, recurrent subconjunctival hemorrhages and conjunctival irritation with chemosis and foreign body sensation [7]. Orbital and ocular adnexa alterations such as blepharochalasis, palpebral ptosis, floppy eyelid, ectropion, hypertelorism, epicanthal folds, orbital pain and periorbital hemorrhages or swelling have been described [8–10]. Moreover, in EDS, tear film instability and deficiency, blue sclera, myopia and degenerative myopia have been reported, oftentimes associated with vitreo-retinal degenerations and abnormalities due to scleral thinning and bulging in consequence of abnormal collagen deposition. High myopia can lead to the development of myopic staphyloma, choroidal thinning, retinal atrophy and lacquer cracks [11,12]. Strabismus was reported in this clinical condition, mostly due to craniofacial bone asymmetry with orbital misalignment, and to abnormal collagen deposition in the extraocular muscles. Ocular imbalance due to orbital discrepancy can also cause diplopia, amblyopia and stereopsis reduction [7,13,14]. A higher incidence of open angle glaucoma associated with altered development of the aqueous outflow tract has been reported. This complication is mostly detected in vEDS, because collagen type III is a fundamental component of the juxtacanalicular meshwork and Schlemms canal [15]. Regarding retinal involvement, this condition is characterized by vessel fragility and malformation that may lead to vitreous hemorrhages. In particular, vEDS is linked to a higher incidence of rhegmatogenous retinal detachment and angioid streak formation [16]. Moreover, EDS patients suffering from cardiovascular and valvular diseases are more likely to develop retinal ischemic perivascular lesions, recognizable with ocular coherence tomography [17]. Lens involvement occurs less frequently compared to other heritable connective tissue disorders, particularly Marfan Syndrome. Ectopia lentis and dislocation have rarely been described [18,19]. Interestingly, a higher incidence of small lens opacities has been found among young EDS patients [12]. Furthermore, in patients with important scleral thinning, cases of spontaneous globe rupture have been described. A secondary cause of vision loss can be linked to systemic and cerebrovascular complications, such as cervical artery dissection or aneurysms, which can determine ischemic/hemorrhagic brain injury. These events can manifest with visual symptoms such as nystagmus, visual field loss, visual acuity reduction, diplopia and ocular motility disorders. Carotid cavernous fistula has also been described in EDS patients, potentially responsible for relative afferent pupillary defects, episcleral vein congestion, pulsating exophthalmos and retinal thromboembolism [7]. The most recent EDS classification includes the brittle cornea syndrome. This pathology has an autosomal recessive inheritance and patients present mostly ocular signs. Pathognomonic characteristics of this syndrome are a thin cornea associated with an elevated risk of developing keratoglobus or keratoconus in early life. This condition can cause corneal rupture, perforation and infections as consequences of mild injuries in young patients. Progressive corneal thinning may result in corneal opacities and scarring, potentially vision-threatening. Central corneal thickness is often <400 μm. These patients are prone to high myopia due to an elevated ocular axial length, associated with blue sclera. In these cases, patients can also be subjected to retinal complications such as retinal detachment or secondary glaucoma. Additionally, this pathology presents a systemic involvement with deafness, hypercompliant tympanic membrane, hip dysplasia, scoliosis, arachnodactyly, joint hypermobility, finger contractures and hyper-extensible and fragile skin [3,7,20]. Moreover, in BCS patients, altered ECM protein expression in Bruch's membrane has been demonstrated. Bruch's membrane weakness can be responsible for the development of choroidal neovascularization at a young age [21].

In this cross-sectional observational study, we evaluated the orthoptic features and the ocular motility (OM) abnormalities associated with Ehlers-Danlos Syndrome.

2. Materials and Methods

From January 2021 to December 2022, 132 patients with EDS were consecutively observed from the Centre of Rare Diseases at the Policlinico Umberto I University Hospital. Among these 132 patients, 41 patients were recruited for the study, which was approved by Sapienza University of Rome's Ethics Board and was performed in accordance with the Declaration of Helsinki. Written informed consent was obtained from all participants. The 41 enrolled patients were 36 females and 5 males, with a mean age of 28 + 12.6 (range 6–55). Moreover, 90.2% of patients (N = 37) had hypermobile EDS, 2.4% (N = 1) a vascular type and 7.3% (N = 3) of patients were not classified. All patients were already diagnosed with EDS and genetically classified. A full orthoptic evaluation, including the corneal light reflex test (CLR) and the cover test (CT) examination, was performed. We checked for the presence of OM disorders and associated diplopia or tenderness during ocular movements in all gaze positions. The stereoscopic sense was evaluated with the Lang test I and II. The BCVA evaluation at 5 m with the Snellen chart and a full ophthalmological examination were performed only during the recruitment stage, with the sole purpose of screening the patients for the exclusion criteria. The exclusion criteria were all ophthalmological diseases apart from OM abnormalities, and a Best Corrected Visual Acuity (BCVA) below a 0.1 decimal, to guarantee correct fixation during the orthoptic examination. Moreover, patients with vascular, neurodegenerative and systemic diseases (e.g., multiple sclerosis, Parkinson's disease or diabetes) that could influence the orthoptic evaluation were excluded.

Statistical Analysis

Descriptive statistical analysis was performed considering all data collected. For each categorical variable, percentages and frequencies were calculated. The chi square test was employed to evaluate possible associations between the orthoptic parameters. The analysis was conducted using R (version 4.2.2, R Foundation for Statistical Computing, Vienna, Austria). A p value < 0.05 was considered significant.

3. Results

Orthoptic Evaluation

The CLR at 33 cm appeared symmetrical in 58.5% of patients (N = 24). Moreover, this test showed an intermittent divergent deviation in 31.7% (N = 13), an intermittent horizontal deviation associated with a vertical deviation in 4.9% (N = 2), a manifest strabismus in 2.4% (N = 1) and a microstrabismus in 2.4% (N = 1) of patients.

The CLR at 5 m appeared symmetrical in 85.4% (N = 35) of patients. This examination showed an intermittent exotropia in 2.4% (N = 1), a manifest strabismus in 2.4% (N = 1) and a microstrabismus in 9.8% (N = 4) of patients.

The CT at 33 cm revealed orthophoria in 19.5% (N = 8), exophoria in 22% (N = 9), intermittent exotropia in 43.9% (N = 18), exotropia in 2.4% (N = 1), esophoria in 2.4% (N = 1), intermittent esotropia in 4.9% (N = 2) and esotropia in 4.9% (N = 2) of patients.

The CT at 5 m revealed orthophoria in 75.6% (N = 31), exophoria in 9.8% (N = 4), intermittent exotropia in 7.3 % (N = 3), esophoria in 2.4% (N = 1) and esotropia in 4.9% (N = 2) of patients.

The CT at 33 cm showed a vertical deviation in 65.9% (N = 27) of patients; in particular, 41.5% (N = 17) presented a right/left (R/L) and 24.4% (N = 10) a left/right (L/R) deviation. The CT at 5 m showed a vertical deviation in 24.4% (N=10); in particular, R/L was observed in 14.6% (N = 6) and L/R in 9.8% (N = 4) of patients.

The Lang test showed an absence of stereopsis in 4.9% (N = 2) of patients. One of the patients revealed a microstrabismus and the other one presented a partially accommodative esotropia with an angle of 20 prismatic diopters (PD) measured at 33 cm and 40 PD measured at 5 m.

All data regarding OM of the right eye (RE) and the left eye (LE) are summarized in Tables 1 and 2.

Table 1. Ocular motility of the right eye.

Muscle	Severe Muscle Hyperfunction	Moderate–Mild Hyperfunction	Severe Muscle Hypofunction	Moderate–Mild Hypofunction	Normal Muscle Functioning
MR	2.4% (N = 1)	12.2% (N = 5)	14.6% (N = 6)	29% (N = 12)	41.4% (N = 17)
LR	17.1% (N = 7)	26.8% (N = 11)	4.9% (N = 2)	2.4% (N = 1)	48.7% (N = 20)
SR	N = 0	N = 0	17.1% (N = 7)	24.4% (N = 10)	58.5% (N = 24)
IR	N = 0	N = 0	N = 0	N = 0	100% (N = 41)
SO	N = 0	2.4% (N = 1)	N = 0	N = 0	97.6% (N = 40)
IO	24.4% (N = 10)	34.2% (N = 14)	N = 0	N = 0	41.5% (N = 17)

MR: Medial Rectus, LR: Lateral rectus, SR: Superior Rectus, IR: Inferior Rectus, SO: Superior Oblique, IO: Inferior Oblique.

Table 2. Ocular motility of the left eye.

Muscle	Severe Muscle Hyperfunction	Moderate–Mild Hypofunction	Severe Muscle Hypofunction	Moderate–Mild Hypofunction	Normal Muscle Functioning
MR	4.9% (N = 2)	7.3% (N = 3)	19.5% (N = 8)	29.2% (N = 12)	39% (N = 16)
LR	9.8% (N = 4)	28.4% (N = 12)	2.4% (N = 1)	2.4% (N = 1)	56% (N = 23)
SR	N = 0	2.4% (N = 1)	17.1% (N = 7)	36.8% (N = 11)	53.7% (N = 22)
IR	N = 0	N = 0	N = 0	N = 0	100% (N = 41)
SO	N = 0	N = 0	N = 0	N = 0	100% (N = 41)
IO	9.8 % (N = 4)	19.5 % (N = 8)	N = 0	N = 0	70.7% (N = 29)

MR: Medial Rectus, LR: Lateral rectus, SR: Superior Rectus, IR: Inferior Rectus, SO: Superior Oblique, IO: Inferior Oblique.

Among our patients, 41.5% (N = 17) complained of diplopia, while 58.5% (N = 24) showed single binocular vision. Furthermore, 75.6% (N = 31) of patients reported tenderness during OM examination. Half of the diplopic patients presented pain in the upward gaze and in the right or left lateroversion. The inferential analysis, reported in Tables 3 and 4, allowed us to evaluate the associations between the orthoptic parameters and the presence of diplopia and pain. Significant associations were observed between a Medial Rectus muscle (MR) deficit of the RE with pain (p = 0.020) and diplopia (p = 0.014). Furthermore, there was also a significant association between alteration of the Lateral Rectus muscle (LR) of the RE and the presence of eye pain (p = 0.004). As regards the LE, a significant association was observed between the MR deficit and pain (p < 0.001) and diplopia (p = 0.030) and between the LR deficit and pain (p = 0.011).

Table 3. Univariate analysis of the RE orthoptic parameters with pain/diplopia.

Variable		Pain		p	Diplopia		p
		Absent	Present		Absent	Present	
MR	Normal	9	10	0.020 °	15	4	0.014 *
	Alteration	1	21		9	13	
LR	Normal	9	11	0.004 °	14	6	0.146 *
	Alteration	1	20		10	11	
SR	Normal	8	16	0.152 °	16	8	0.209 *
	Alteration	2	15		8	9	
IR	Normal	10	31	NC	24	17	NC
	Alteration	—	—		—	—	
SO	Normal	10	30	1.000 °	24	16	0.415 °
	Alteration	0	1		0	1	
IO	Normal	7	10	0.063 °	12	5	0.187 *
	Alteration	3	21		12	12	

MR: Medial Rectus, LR: Lateral rectus, SR: Superior Rectus, IR: Inferior Rectus, SO: Superior Oblique, IO: Inferior Oblique. * p-value Chi-Square Test; ° p-value Fisher Test; NC: not calculable.

Table 4. Univariate analysis of the LE orthoptic parameters with pain/diplopia.

Variable		Pain		p	Diplopia		p
		Absent	Present		Absent	Present	
MR	Normal	9	7	<0.001 °	14	2	0.030 *
	Alteration	1	24		10	15	
LR	Normal	9	13	0.011 °	14	8	0.476 *
	Alteration	1	18		10	9	
SR	Normal	7	15	0.292 °	15	7	0.177 *
	Alteration	3	16		9	10	
IR	Normal	10	31	NC	24	17	NC
	Alteration	—	—		—	—	
SO	Normal	10	31	NC	24	17	NC
	Alteration	—	—		—	—	
IO	Normal	8	21	0.694 °	19	10	0.184 °
	Alteration	2	10		5	7	

MR: Medial Rectus, LR: Lateral rectus, SR: Superior Rectus, IR: Inferior Rectus, SO: Superior Oblique, IO: Inferior Oblique. * p-value Chi-Square Test; ° p-value Fisher Test; NC: not calculable.

4. Discussion

In our study, a complete and detailed OM evaluation describes the characteristics of the extraocular muscle disorders in patients with diagnosed EDS. A manifest strabismus was observed in a small group of the included patients. However, several extraocular muscles, such as MR, LR, SR and IO, were affected, with different frequencies, by hypofunction or hyperfunction, while the IR and SO showed no alteration. In detail, considering the diagnostic gaze position in the right lateroversion, the left MR had severe hypofunction in 19.5% and the right LR had moderate–mild hyperfunction in 26.8% of cases. On the other hand, in the left lateroversion, moderate–mild hypofunction was observed in the right MR in 29% and hyperfunction of the same degree in the left LR in 28.4% of cases. Considering the diagnostic gaze position up and to the right, we found moderate–mild hypofunction in the right SR in 24.4%, with contralateral synergistic muscle hyperfunction (left IO) in 19.5% of cases. The last gaze position in which the OM examination revealed deficits was up and to the left, where the left SR showed moderate–mild hypofunction in 36.8% and the right IO showed moderate–mild hyperfunction in 34.2% of cases. The two horizontal gaze positions to the right and left were associated with the presence of eye pain and diplopia. The presence of diplopia shows how oculomotor alteration appeared after the plastic period (0–6 years) as a consequence of the EDS progression. These orthoptic alterations suggest that the ligamentous laxity observed in EDS tends to involve the extraocular muscles. Our findings suggest that ligamentous laxity and muscular fiber alterations can in fact affect the ocular district, determining functional changes over the years. Ophthalmological and orthoptic evaluation should be considered as part of the regular diagnostic assessment in this pathology. In our cohort, only a minority of patients presenting ocular motility disorders reported diplopia. These data suggest that an important part of ocular muscular dysfunction remains subclinical and underdiagnosed. An early diagnosis of this muscular involvement might be crucial in these patients to monitor the evolution of the disease over time. Moreover, the ocular muscular district offers the possibility of an easy clinical evaluation and can mirror the condition of the systemic muscular apparatus. Furthermore, EDS children during the plastic period should be closely checked, because the development of subclinical strabismus can represent an important cause of amblyopia. In children, the occurrence of subtle forms of strabismus is not associated with the onset of double vision, but can lead to silent sight suppression.

Meyer et al. studied the structures of the fibrils of the dermal collagen and the fibrils of the extraocular muscles and the conjunctivae of both eyes of a child affected by EDS. They observed that in the reticular dermis, 48% of fibrils presented a normal diameter, 23% were enlarged and 29% were thinner than normal. In the extraocular muscle, they found that 77% of fibrils were of normal size, 14.5% of larger size and 8.5% of smaller size. In the conjunctivae, 73% of the fibrils were of normal size, 22% were enlarged and 5% were smaller. In healthy controls, no small fibrils were found in the extraocular muscles [14]. To the best of our knowledge, this is the first study that analyzes specifically the extraocular muscle abnormalities in EDS through a complete and accurate orthoptic assessment. Furthermore, the previous studies that described OM alterations in EDS had by far a smaller sample compared to ours. Perez-Roustit et al. reported OM disorders in 15 (71.4%) of 21 patients affected by EDS, with convergence insufficiency in 13 of them [22]. Louie et al. reported, among 467 patients with confirmed EDS, 17 cases that underwent strabismus correction surgeries, out of which 14 (82.4%) underwent surgery before receiving their EDS diagnosis [13]. Other authors reported strabismus in EDS patients in combination with particular craniofacial features such as down-slanting palpebral fissures, palatine alterations, microretrognathia or protruding jaw and crowded teeth, contributing to congenital contractures and malocclusion [23,24]. It can be hypothesized that bone craniofacial alterations, malocclusion and frequent orthopedic spine issues, leading to incorrect posture, might represent an additional risk factor for the development of strabismus [25,26]. Moreover, we can suppose that these OM disorders are additionally attributable to muscular dysfunction due to the altered deposition and organization of collagen fibrils in the muscles, leading to reduced muscle mass and weakness. Muscular atrophy, hypotonia and contractures are described in the literature in this pathology relatively to other muscular districts, defining a myopathic subtype of EDS [27]. In our sample, these alterations in OM cannot be correlated with compromised binocular vision, because motor fusion was within the normal limits and stereopsis, although in some cases coarse, was present. Promising results in the assessment of stereopsis could be obtained by measuring the ocular following responses (OFR). OFR are configured as short-latency, slow eye movements that constitute a visual tracking system. These responses aim to adjust the fixation as the visual stimulus varies. OFR allows physiological shifts in fixation, between near and far objects appearing in the visual field during everyday activities. Binocular gaze shift is usually permitted thanks to a combination of saccades and vergence movements. It helps to stabilize the eyes on the visual scene [28,29]. Different studies highlighted a possible role of OFR in the evaluation of binocular vision and inter-eye collaboration, through the analysis of eye movements. Neurons localized in the primary visual cortex elicit stronger responses if activated by binocular stimuli, determining enhanced OFR. These findings suggest that OFR might reveal the presence of binocular collaboration and summation in patients [30]. Moreover, OFR can be detected also in young children or in patients in whom it may be difficult to assess stereopsis using common tests for binocular vision [31]. The present study has some limitations, such as its retrospective nature, the small sample size and the poor homogeneity of the sample relative to gender and age, which could affect the generalizability of the findings. On the other hand, as EDS is an extremely rare disease, we chose to recruit, in our study, the largest number of affected patients that we had at our disposal, despite the fact that this has disadvantaged data homogeneity. Another important limit of our study is that our main focus has been on ocular motility disorders, without taking into consideration the possibility to correlate these data with other eventual intraocular manifestations. Further studies are warranted to evaluate the potential correlation between alterations in eye movement and muscular alterations in other corporeal districts. Another field in which to improve research on EDS could be represented by the histological analysis of extraocular muscle features. At present, only one article on this topic is available in the literature [14].

5. Conclusions

The present study describes the characteristics of OM abnormalities in patients with EDS and suggests how accurate orthoptic screening together with the ophthalmological evaluation could provide a significant contribution to the management of the patients affected by this rare syndrome. An orthoptic protocol of rehabilitation should be planned in cases of OM alterations that interfere with the patient's quality of life. Further studies are warranted to confirm the findings of the present research on the OM abnormalities associated with EDS, and it would be desirable to perform a genetic investigation to understand the link between the gene and the extraocular muscle involved.

Author Contributions: Conceptualization, A.M.C., L.I., M.C. and A.Z.; data gathering and analysis, A.M.C., L.I., M.A., M.G. (Marika Graziani) and L.C.; writing—original draft preparation, A.M.C., L.I., M.G. (Marika Graziani), R.M. and A.Z.; editing, L.I., M.A., M.G. (Magda Gharbiya) and A.L.; supervision, M.C., M.G. (Magda Gharbiya) and A.L. All authors have read and agreed to the published version of the manuscript.

Funding: This research received no external funding.

Institutional Review Board Statement: The study was conducted in accordance with the Declaration of Helsinki, and approved by the Ethics Committee of AOU Policlinico Umberto I- Sapienza Università di Roma (protocol code 0855/2020 and date of approval 14/09/2020).

Informed Consent Statement: Informed consent was obtained from all subjects involved in the study.

Data Availability Statement: The data that support the findings of this study are not publicly available to protect the privacy of research participants but are available from the corresponding author, L.I.

Conflicts of Interest: The authors declare no conflict of interest.

References

1. Malfait, F.; Castori, M.; Francomano, C.A.; Giunta, C.; Kosho, T.; Byers, P.H. The Ehlers-Danlos syndromes. *Nat. Rev. Dis. Primers* **2020**, *6*, 64. [CrossRef]
2. Beighton, P.; De Paepe, A.; Steinmann, B.; Tsipouras, P.; Wenstrup, R.J. Ehlers-Danlos syndromes: Revised nosology, Villefranche, 1997. Ehlers-Danlos National Foundation (USA) and Ehlers-Danlos Support Group (UK). *Am. J. Med. Genet.* **1998**, *77*, 31–37. [CrossRef]
3. Malfait, F.; Francomano, C.; Byers, P.; Belmont, J.; Berglund, B.; Black, B.; Bloom, L.; Bowen, J.M.; Brady, A.F.; Burrows, N.P.; et al. The 2017 international classification of the Ehlers-Danlos syndromes. *Am. J. Med. Genet. Part C Semin. Med. Genet.* **2017**, *175*, 8–26. [CrossRef] [PubMed]
4. Meester, J.A.N.; Verstraeten, A.; Schepers, D.; Alaerts, M.; Van Laer, L.; Loeys, B.L. Differences in manifestations of Marfan syndrome, Ehlers-Danlos syndrome, and Loeys-Dietz syndrome. *Ann. Cardiothorac. Surg.* **2017**, *6*, 582–594. [CrossRef]
5. Ritelli, M.; Colombi, M. Molecular Genetics and Pathogenesis of Ehlers-Danlos Syndrome and Related Connective Tissue Disorders. *Genes* **2020**, *11*, 547. [CrossRef]
6. McDermott, M.L.; Holladay, J.; Liu, D.; Puklin, J.E.; Shin, D.H.; Cowden, J.W. Corneal topography in Ehlers-Danlos syndrome. *J. Cataract. Refract. Surg.* **1998**, *24*, 1212–1215. [CrossRef]
7. Asanad, S.; Bayomi, M.; Brown, D.; Buzzard, J.; Lai, E.; Ling, C.; Miglani, T.; Mohammed, T.; Tsai, J.; Uddin, O.; et al. Ehlers-Danlos syndromes and their manifestations in the visual system. *Front. Med.* **2022**, *9*, 996458. [CrossRef] [PubMed]
8. Segev, F.; Héon, E.; Cole, W.G.; Wenstrup, R.J.; Young, F.; Slomovic, A.R.; Rootman, D.S.; Whitaker-Menezes, D.; Chervoneva, I.; Birk, D.E. Structural Abnormalities of the Cornea and Lid Resulting from Collagen V Mutations. *Investig. Opthalmol. Vis. Sci.* **2006**, *47*, 565–573. [CrossRef]
9. Ritelli, M.; Venturini, M.; Cinquina, V.; Chiarelli, N.; Colombi, M. Multisystemic manifestations in a cohort of 75 classical Ehlers-Danlos syndrome patients: Natural history and nosological perspectives. *Orphanet J. Rare Dis.* **2020**, *15*, 197. [CrossRef]
10. Colombi, M.; Dordoni, C.; Venturini, M.; Ciaccio, C.; Morlino, S.; Chiarelli, N.; Zanca, A.; Calzavara-Pinton, P.; Zoppi, N.; Castori, M.; et al. Spectrum of mucocutaneous, ocular and facial features and delineation of novel presentations in 62 classical Ehlers-Danlos syndrome patients. *Clin. Genet.* **2017**, *92*, 624–631. [CrossRef]
11. Brady, A.F.; Demirdas, S.; Fournel-Gigleux, S.; Ghali, N.; Giunta, C.; Kapferer-Seebacher, I.; Kosho, T.; Mendoza-Londono, R.; Pope, M.F.; Rohrbach, M.; et al. The Ehlers-Danlos syndromes, rare types. *Am. J. Med Genet. Part C Semin. Med Genet.* **2017**, *175*, 70–115. [CrossRef] [PubMed]
12. Gharbiya, M.; Moramarco, A.; Castori, M.; Parisi, F.; Celletti, C.; Marenco, M.; Mariani, I.; Grammatico, P.; Camerota, F. Ocular Features in Joint Hypermobility Syndrome/Ehlers-Danlos Syndrome Hypermobility Type: A Clinical and In Vivo Confocal Microscopy Study. *Am. J. Ophthalmol.* **2012**, *154*, 593–600.e1. [CrossRef] [PubMed]
13. Louie, A.; Meyerle, C.; Francomano, C.; Srikumaran, D.; Merali, F.; Doyle, J.J.; Bower, K.; Bloom, L.; Boland, M.V.; Mahoney, N.; et al. Survey of Ehlers–Danlos Patients' ophthalmic surgery experiences. *Mol. Genet. Genom. Med.* **2020**, *8*, e1155. [CrossRef] [PubMed]
14. Meyer, E.; Ludatscher, R.M.; Zonis, S. Collagen Fibril Abnormalities in the Extraocular Muscles in Ehlers-Danlos Syndrome. *J. Pediatr. Ophthalmol. Strabismus* **1988**, *25*, 67–72. [CrossRef] [PubMed]
15. Mitra, A.; Ramakrishnan, R.; Kader, M.A. Open angle glaucoma in a case of Type IV Ehler Danlos syndrome: A rarely reported association. *Indian J. Ophthalmol.* **2014**, *62*, 880–884. [CrossRef]
16. Chikamoto, N.; Teranishi, S.; Chikama, T.-I.; Nishida, T.; Ohshima, K.; Hatsukawa, Y. Abnormal Retinal Blood Vessels in Ehlers-Danlos Syndrome Type VI. *Jpn. J. Ophthalmol.* **2007**, *51*, 453–455. [CrossRef]
17. Long, C.P.; Chan, A.X.; Bakhoum, C.Y.; Toomey, C.B.; Madala, S.; Garg, A.K.; Freeman, W.R.; Goldbaum, M.H.; DeMaria, A.N.; Bakhoum, M.F. Prevalence of subclinical retinal ischemia in patients with cardiovascular disease—A hypothesis driven study. *Eclinicalmedicine* **2021**, *33*, 100775. [CrossRef]
18. Nelson, L.B.; Maumenee, I.H. Ectopia lentis. *Surv. Ophthalmol.* **1982**, *27*, 143–160. [CrossRef]
19. Klaassens, M.; Reinstein, E.; Hilhorst-Hofstee, Y.; Schrander, J.; Malfait, F.; Staal, H.; Have, L.T.; Blaauw, J.; Roggeveen, H.; Krakow, D.; et al. Ehlers-Danlos arthrochalasia type (VIIA-B)—Expanding the phenotype: From prenatal life through adulthood. *Clin. Genet.* **2012**, *82*, 121–130. [CrossRef]
20. Walkden, A.; Wright, E.B.; Au, L. Brittle cornea syndrome: Current perspectives. *Clin. Ophthalmol.* **2019**, *13*, 1511–1516. [CrossRef]
21. Porter, L.F.; Gallego-Pinazo, R.; Keeling, C.L.; Kamieniorz, M.; Zoppi, N.; Colombi, M.; Giunta, C.; Bonshek, R.; Manson, F.D.; Black, G.C. Bruch's membrane abnormalities in PRDM5-related brittle cornea syndrome. *Orphanet J. Rare Dis.* **2015**, *10*, 145. [CrossRef]
22. Perez-Roustit, S.; Nguyen, D.-T.; Xerri, O.; Robert, M.-P.; De Vergnes, N.; Mincheva, Z.; Benistan, K.; Bremond-Gignac, D. Ocular manifestations in Ehlers-Danlos Syndromes: Clinical study of 21 patients. *J. Fr. Ophtalmol.* **2019**, *42*, 722–729. [CrossRef]
23. Kosho, T.; Miyake, N.; Hatamochi, A.; Takahashi, J.; Kato, H.; Miyahara, T.; Igawa, Y.; Yasui, H.; Ishida, T.; Ono, K.; et al. A new Ehlers-Danlos syndrome with craniofacial characteristics, multiple congenital contractures, progressive joint and skin laxity, and multisystem fragility-related manifestations. *Am. J. Med Genet. Part A* **2010**, *152A*, 1333–1346. [CrossRef] [PubMed]
24. Stappert, D.; Bakhsh, A.; Wiese, L. Class II malocclusion and ectopic canine treated with Pendulum appliance and TADs in a high-relapse potential Ehlers-Danlos patient: A case report. *J. World Fed. Orthod.* **2020**, *9*, 25–31. [CrossRef]

25. Silvestrini-Biavati, A.; Migliorati, M.; Demarziani, E.; Tecco, S.; Silvestrini-Biavati, P.; Polimeni, A.; Saccucci, M. Clinical association between teeth malocclusions, wrong posture and ocular convergence disorders: An epidemiological investigation on primary school children. *BMC Pediatr.* **2013**, *13*, 12. [CrossRef]
26. Marathe, N.; Lohkamp, L.-N.; Fehlings, M.G. Spinal manifestations of Ehlers-Danlos syndrome: A scoping review. *J. Neurosurg. Spine* **2022**, *37*, 783–793. [CrossRef]
27. Delbaere, S.; Dhooge, T.; Syx, D.; Petit, F.; Goemans, N.; Destree, A.; Vanakker, O.; De Rycke, R.; Symoens, S.; Malfait, F.; et al. Novel defects in collagen XII and VI expand the mixed myopathy/Ehlers-Danlos syndrome spectrum and lead to variant-specific alterations in the extracellular matrix. *Genet Med.* **2020**, *22*, 112–123. [CrossRef] [PubMed]
28. Takemura, A.; Kawano, K. Sensory-to-motor processing of the ocular-following response. *Neurosci Res.* **2002**, *43*, 201–206. [CrossRef]
29. Joshi, A.C.; Thurtell, M.J.; Walker, M.F.; Serra, A.; Leigh, R.J. Effect of vergence on human ocular following response (OFR). *J. Neurophysiol.* **2009**, *102*, 513–522. [CrossRef] [PubMed]
30. Quaia, C.; Optican, L.M.; Cumming, B.G. Binocular summation for reflexive eye movements. *J. Vis.* **2018**, *18*, 7. [CrossRef] [PubMed]
31. Miladinović, A.; Quaia, C.; Ajčević, M.; Diplotti, L.; Cumming, B.G.; Pensiero, S.; Accardo, A. Ocular-following responses in school-age children. *PLoS ONE* **2022**, *17*, e0277443. [CrossRef] [PubMed]

applied sciences

MDPI

Review

Diagnosis, Follow-Up and Therapy for Secondary Osteoporosis in Vulnerable Children: A Narrative Review

Anne T. M. Dittrich [1], Etienne J. M. Janssen [1,2], Joyce Geelen [1], Karlijn Bouman [3], Leanne M. Ward [4,5] and Jos M. T. Draaisma [1,*]

1 Department of Pediatrics, Radboudumc Amalia Children's Hospital, Radboud Institute for Health Sciences, Radboud University Medical Center, 6525 GA Nijmegen, The Netherlands
2 Department of Pediatrics, Maastricht University Medical Centre, 6229 HX Maastricht, The Netherlands
3 Department of Pediatric Neurology, Donders Institute for Brain, Cognition and Behavior, Amalia Children's Hospital, Radboud University Medical Center, 6525 GA Nijmegen, The Netherlands
4 The Department of Pediatrics, University of Ottawa, Ottawa, ON K1H 8L1, Canada
5 The Ottawa Pediatric Bone Health Research Group, Children's Hospital of Eastern Ontario Research Institute, Ottawa, ON K1H 5B2, Canada
* Correspondence: jos.draaisma@radboudumc.nl

Abstract: By definition, children constitute a vulnerable population, especially when they are chronically ill and/or disabled. A characteristic of chronically ill and disabled children is that they also suffer from indirect effects of their disease, such as immobilization, chronic inflammation, reduced time outdoors in the sun, osteotoxic effects of disease-targeted therapy (like glucocorticoids), and poor nutrition. All these factors may lead to bone fragility due to secondary osteoporosis, a co-morbidity that may be overlooked in the context of serious underlying diseases. The ultimate goal of osteoporosis diagnosis and monitoring in this setting is the early identification, prevention, and treatment of low-trauma long bone and vertebral fractures; indeed, vertebral fractures are a frequently under-diagnosed manifestation of overt bone fragility in this context. Efforts to prevent first-ever fractures are also meritorious, including encouragement of weight-bearing activities, optimization of nutritional status, including calcium and vitamin D supplementation, and the diagnosis and treatment of delayed growth and puberty; however, these conservative measures may be insufficient in those at high risk. Numerous natural history studies have shown that vertebral fractures are more common than non-vertebral (i.e., long bone) fractures in at-risk children. Not surprisingly, the cornerstone of secondary osteoporosis monitoring is lateral spine imaging for the early detection of vertebral collapse. Although dual-energy x-ray absorptiometry (DXA) is the gold standard to measure bone mineral density, digital X-ray radiogrammetry may be used as a surrogate measure of bone strength if dual-energy x-ray absorptiometry is not available. In the event that preventive measures fail, treatment with bisphosphonates may be appropriate. Typically, treatment with intravenous bisphosphonates is reserved for children with overt bone fragility and limited potential for spontaneous recovery. However, there is increasing attention to very high-risk children, such as boys with Duchenne muscular dystrophy, who may benefit from bisphosphonate therapy prior to first-ever fractures (given their high fracture frequency and essentially absent potential for spontaneous recovery). This article provides a contemporary overview of the definition and diagnosis of osteoporosis in children with chronic illness, along with the approach to monitoring those at risk and the evidence for currently recommended intervention strategies.

Keywords: secondary osteoporosis; prevalence; prevention; screening; therapy

check for updates

Citation: Dittrich, A.T.M.; Janssen, E.J.M.; Geelen, J.; Bouman, K.; Ward, L.M.; Draaisma, J.M.T. Diagnosis, Follow-Up and Therapy for Secondary Osteoporosis in Vulnerable Children: A Narrative Review. *Appl. Sci.* **2023**, *13*, 4491. https://doi.org/10.3390/app13074491

Academic Editors: Gaspare Palaia, Alice Bruscolini and Massimo Ralli

Received: 17 January 2023
Revised: 6 March 2023
Accepted: 29 March 2023
Published: 1 April 2023

1. Introduction

The health domains of vulnerable populations can be divided into physical, psychological, and social [1]. The physical domain includes all infants and chronically ill and disabled children and adults. Most people in this domain have more than two chronic morbidities by the age of 65 [2]. Osteoporosis is a known chronic morbidity in adults by the age of 65 [3]. In children, osteoporosis can be categorized into primary and secondary causes. Primary osteoporosis in children is caused by a disease of the bone itself, for example, osteogenesis imperfecta. Secondary osteoporosis in children can be caused by immobilization, cytokine-release-related disorders, gastrointestinal disorders, endocrine disorders, drugs (particularly glucocorticoids), or a combination of these factors. In children with certain diseases, multiple factors can contribute to bone fragility and secondary osteoporosis; the most common of these include chronic glucocorticoid use, inflammatory disorders, and compromised mobility. Both types of osteoporosis are associated with an increased risk of fractures and morbidity and decreased quality of life [4].

The diagnosis of osteoporosis in children is more challenging than in adults because bone mass varies greatly with age/pubertal stage and because peak bone mass has not yet been attained. In addition, fracture rates increase during periods of accelerated linear growth, such as during adolescence [5]. In 2014, The International Society of Clinical Densitometry (ISCD) put forward a definition of osteoporosis that was meant to avoid over-diagnosis in otherwise healthy children. In this specific context, the ISCD defined pediatric osteoporosis as 1. $\geq$2 long bone fractures before the age of 10 years or $\geq$3 long bone fractures before the age of 19 years in combination with a low bone mineral density (BMD) for age and gender (z-score ≤ -2.0 as measured by DXA), or 2. $\geq$1 vertebral compression fractures (VF), independent of the BMD, in the absence of major trauma or local disease [6]. It is important to underscore that this conservative definition was intended for otherwise healthy children, as opposed to children at high risk for fractures, so as to avoid labeling children with osteoporosis who may have simply been unlucky on the playground or in sports activities. In children with significant risk factors for bone fragility due to chronic illness, even a single low-trauma long bone or vertebral fracture (in the absence of a "low BMD") may be sufficient to diagnose the child with osteoporosis, as discussed in a recent paper by Ward et al. [7].

This article will first discuss the diagnostic approach to bone fragility and low bone mass in childhood. We will then focus on two representative conditions associated with low bone mass: immobilization and drug-related (glucocorticoid therapy). Thereafter, we will describe methods that can be used for osteoporosis screening, prevention, and therapy. Finally, we will describe secondary osteoporosis in three groups of children with chronic diseases, to which less attention is paid in literature, more in detail: children with profound intellectual and multiple disabilities, children on the ketogenic diet, and children with congenital myopathies.

2. Diagnosis

As stated earlier, the ISCD definition of osteoporosis in children was not intended for children with significant risk factors for bone fragility due to chronic illness. Waiting for a second long bone fracture, a first vertebral fracture or a low BMD by DXA following low-trauma fractures unnecessarily delays the initiation of treatment in those with conditions known to be associated with an increased risk of bone fragility [8]. A vertebral fracture in children has been defined as >20% loss of vertebral height ratio according to the semi-quantitative Genant method [9]. The Genant method to define VF in children has been validated by showing that vertebral fractures defined in this way independently predict new (i.e., subsequent) vertebral and long bone fractures [8]. VFs are more common than long bone fractures in children with chronic illness treated with glucocorticoids: VFs occur in 32.5%, non-VFs in 23% of children with acute lymphoblastic leukemia [8], and VFs occur in 16.3% and non-VF fractures in 10.1% of children with rheumatic disorders treated with glucocorticoids [10]. Thirty-nine percent of children with VF in the group of children with acute lymphoblastic leukemia were asymptomatic [8]. For this reason, routine monitoring by lateral spine imaging (radiography or dual-energy x-ray absorptiometry) is required for their early detection [11]. Untreated VFs can lead to chronic back pain and spine deformity [8]. Therefore, screening for VF is so important that many clinicians are now using DXA-based Vertebral Fracture Assessment, for which there are guidelines in children put forward by the ISCD [12,13]. This has the advantage of low radiation, lack of parallax, and showing the entire spine on one cassette (which minimizes errors in determining the affected vertebral level(s)) [12].

A child with a chronic illness associated with bone fragility with a size-adjusted BMD z-score <-2.0 but without a history of fractures is not classified as having osteoporosis per se but would be considered at-risk. The definition "fractures without major trauma" has been defined as fractures occurring outside of motor vehicle accidents or falling from 10 feet (3 m) or less [4]. With respect to falls in the at-risk, chronic illness setting, a more appropriate definition has been used: falling from a standing height or less or at no more than walking speed [8].

DXA-based BMD can be viewed as a clinical precursor to bone fragility, as it is considered a surrogate for bone strength. Because 90% of peak bone mass is acquired by the age of 18 years, chronic pediatric disease reduces BMD and thus increases fracture risk, not only in childhood and adolescence but also in adulthood [4,14]. There is a strong correlation between BMD and fracture risk [15]. However, BMD can be normal in children with fractures due to secondary osteoporosis [8,15]. DXA is the gold standard for the measurement of bone mass (bone mineral content, BMC) and bone mineral density (BMD) in children because of its precision, minimal radiation dose, reproducibility, availability of normative data, and correlation with prevalent and incident fractures in children [15–17]. Different DXA skeletal sites for BMD measurement have been described, with the lumbar spine (L1-L4] and hip (the latter, in children over 4–5 years of age) representing key skeletal sites [13,16]. The main purpose thus of BMD is to provide additional evidence to justify an appropriate workup and therapeutic intervention. Follow-up of the BMD can be used as a predictor of bone fragility, with a loss of ≥ 0.5 SD considered clinically significant [7,11].

However, DXA has its limitations. DXA may be technically impossible to perform or interpret in children due to movement during measurement, metallic implants, contractures, and sometimes scoliosis [18]. Secondly, the z-scores are based on calendar age and not skeletal maturation. This may lead to inaccurate interpretation of measurements in those with delayed pubertal development [19,20]. Last but not least, DXA provides a measurement of areal BMD (g/cm^2) rather than volumetric density (g/cm^3), which can give an underestimation of BMD in children with small stature s and an overestimation of BMD in children with tall stature [21].

There are other methods, like digital X-ray radiogrammetry of the hand, peripheral quantitative computed tomography (both standard (pQCT) and high resolution (HR-pQCT)), and quantitative ultrasound, that may overcome some of these limitations. However, these are currently not in routine clinical use. A recent systematic review, including a meta-analysis, reviewed the literature on the relationship of digital X-ray radiogrammetry, pQCT, and quantitative ultrasound with DXA [22]. According to this study, digital X-ray radiogrammetry had the strongest positive relationship with DXA (correlation coefficient of 0.71]. Digital X-ray radiogrammetry uses web-based software (for example BoneXpert), which can assess both skeletal maturation (bone age) and bone strength, expressed as the bone health index (BHI). This BHI is a measurement of cortical thickness and cortical mineralization, which results in a representation of bone quality. The BHI reference values are specific for gender and bone age. Digital X-ray radiogrammetry is less stressful for pediatric patients than DXA and is easy to obtain. Often performing an X-ray of the hand does not involve additional exposure to ionizing radiation since hand radiographs for the assessment of bone age are regularly obtained in at-risk children [20]. In a prospective study of 101 pediatric patients with a high probability of low bone strength, digital X-ray radiogrammetry was compared with DXA [18] in children with the following profile: mean age was 11.7 years, 38 were non-ambulatory, and 52 had a neurological disorder. The mean BMD z-score was -1.3 for the group and for the non-ambulatory children -2.2. Digital X-ray radiogrammetry had a sensitivity of 67% and a specificity of 83% for BMD z-score ≤ -2.0. Additionally, If the BMD z-score was <-2.0, digital X-ray radiogrammetry z-scores (being also <-2.0] demonstrated a percentage of 92.4% agreement.

(HR-) pQCT is used to measure cortical and trabecular volumetric bone mineral density separately and microarchitectural bone morphology. Movement artifacts are a real limitation [23]. The use of (HR-) pQCT is limited due to the lack of standardized pediatric reference data for young children, and only a few (HR-) pQTC scanners are available for patient care. The correlation coefficient of (HR-) pQCT with DXA is only 0.57 [21]. Quantitative ultrasound has the same correlation coefficient.

Practical information about who and how to screen for osteoporosis is given in the Section 4 further in this manuscript.

3. Risk Factors for Developing Secondary Osteoporosis

3.1. *Immobilization and Secondary Osteoporosis*

Immobilization is normal in children who are wheelchair-dependent. Cerebral palsy and neuromuscular disorders, such as Duchenne muscular dystrophy, are examples of diseases with immobilization. However, also severe neurodevelopmental disorders and spina bifida are frequent causes of immobilization.

The degree of immobilization in children with cerebral palsy is classified with the Gross Motor Functional Classification Scale (GMFCS) in five levels [24]. GMFCS level 4 means that a child can walk indoors for short distances with assistance but relies on a wheelchair outdoors. Children with GMFCS level 5 are wheelchair dependent. Immobilization in children with cerebral palsy reduces biomechanical bone loading, leading to thinner long bones and less trabecular bone formation [25]. In children aged 2–19 years with moderate to severe cerebral palsy, classified as GMFCS level 3–5, low BMD was found in 97% of children unable to stand and older than 9 years. This leads to reduced periosteal apposition in lower extremity bones, reducing cortical thickness. Consequently, fractures occur most commonly in the distal femur and tibia. Fractures occurred in 26% of children who were older than 10 years [26]. Other factors that contributed to low BMD (z score ≤ -2.0] were feeding difficulty and the use of anticonvulsants. Other studies showed an incidence of fractures of 4% per year [27].

The major role that immobilization plays in secondary osteoporosis and low BMD is also illustrated by boys with Duchenne muscular dystrophy. In the era preceding treatment with glucocorticoids, Larson and Henderson found that BMD was only slightly decreased (z-score lumbar spine -0.8] when the boys were ambulatory but decreased significantly after the loss of ambulation (z-score lumbar spine -1.6] [28]. This was also shown by Crabtree et al. in boys with Duchenne muscular dystrophy treated with glucocorticoids who became non-ambulant [29]. They showed that 44% of the boys sustained a fracture. Two-thirds of these fractures involved the lower extremities, and there were no vertebral fractures. Moreover, 44% of the boys who walked with support at the time of fracture never resumed walking after the fracture. In addition, Joseph et al. showed an absence of clinical vertebral fractures in glucocorticoid naïve boys [30]. In the authors' experience, vertebral fractures can occur in DMD among steroid-naïve patients if routine screening is part of the bone health evaluation; however, this occurrence is rarely related to boys with DMD who are receiving glucocorticoid therapy.

Preclinical studies are useful for a more in-depth understanding of the relationship between immobilization and secondary osteoporosis. Animal models may be used to study immobilization and the cellular mechanisms of secondary osteoporosis. A recent systematic review gives an overview of known animal models [31].

3.2. *Drug-Induced Secondary Osteoporosis*

A myriad of drugs can lead to low BMD and secondary osteoporosis. The most well-known are glucocorticoids, anticonvulsants, and methotrexate [32]. Glucocorticoids are often used for prolonged periods of time in (chronic) diseases in children like systemic inflammatory and autoimmune diseases, renal diseases, after organ transplantation, leukemia, and Duchenne muscular dystrophy. These diseases in themselves may also lead to fragility fractures because of reduced bone strength, for example, due to the effect of the increased cytokines (like IL1, IL6, and tumor necrosis factor-alpha) in case of systemic inflammatory and autoimmune diseases on bone metabolism [33]. Much is known about the adverse effects of glucocorticoids on bone strength [34]. Glucocorticoids cause decreased bone formation, with an additional early and transient increase in bone resorption. The final effect is increased bone turnover with early onset bone loss [35]. BMD rapidly decreases in the first 2 weeks after the start of glucocorticoids, leading to significant bone loss in the first 3–6 months of therapy [36]. This loss diminishes with time and is replaced by a chronic phase of decreased bone formation. The ultimate effect is a reduction of BMD and altered bone microarchitecture, with a predilection for the trabecular-rich spine [34]. This deleterious effect of glucocorticoids on bone strength can be seen in children with Duchenne muscular dystrophy. Glucocorticoids are used to delay loss of ambulation, improve or retain pulmonary function with reduced need for assisted ventilation and delay cardiomyopathy [37]. Before corticosteroids were used, vertebral fractures were rarely seen in children with Duchenne muscular dystrophy, and most fractures involved the lower extremities [28]. In boys living with Duchenne muscular dystrophy who are treated with glucocorticoids, the prevalence of vertebral fractures is >50%, with a cumulative incidence of 28% over a median follow-up of 4 years after starting with glucocorticoids [30,38]. In Canada, an observational cohort study was performed to increase insight into glucocorticoid-induced osteoporosis in children [34]. The most important observations were that vertebral fractures are the hallmark of pediatric glucocorticoid-induced secondary osteoporosis, can occur in the first year of glucocorticoid treatment, and are frequently asymptomatic. However, some children have the growth-mediated ability to restore normal vertebral body dimensions following vertebral fractures. This is important to know since this may preclude the need for intravenous osteoporosis therapy [39]. Children with poor growth, older children (with less residual growth), and children with ongoing bone health threats have less potential for vertebral body reshaping, which can result in permanent vertebral deformity [8]. Therefore, timely intervention with intravenous osteoporosis therapy is paramount in children with vertebral fractures if they have persistent risk factors for ongoing spine collapse.

Although preclinical models may help to understand the effects of glucocorticoids on bone metabolism and bone strength, up to now, there is no robust animal model to evaluate known and new interventions [40].

4. Screening for Secondary Osteoporosis

The goal of screening is to identify patients with a high risk for secondary osteoporosis in order to initiate bone protection therapy in a timely fashion. This has led to screening for early rather than late signs of vertebral fractures, as well as reductions in BMD following appropriate size corrections. This aligns with a secondary prevention approach, which seeks to mitigate the progression of low BMD and bone fragility following identification in an earlier stage [39]. High-risk patients belong to one of the earlier-mentioned groups, especially those with immobilization with or without additional risk factors and those treated with daily or IV glucocorticoids for more than three months.

Recently, two observations have demonstrated the limitations of a "BMD-focused approach" instead of implementing a "fracture- and function-focused approach." First, the use of a BMD z-score threshold of -2 or worse to identify a child as having osteoporosis is problematic due to variability in the z-scores arising from the reference databases [41], and secondly, asymptomatic VF can occur at lumbar spine BMD z-scores >-2, thereby requiring spine imaging for vertebral fractures, especially in children using daily oral or IV glucocorticoids for more than three months where the risk of vertebral fractures is even higher [39]. Other functional outcomes that should also be considered in osteoporosis screening include a history of long bone fractures, growth, pain (especially indicative for vertebral fractures), degree of mobility, muscle strength, and the potential for spontaneous, medication-unassisted recovery (the latter, which is influenced by pubertal stage and residual growth potential). Screening of BMD (by DXA or digital X-ray radiogrammetry) is an adjuvant component of bone health monitoring, as it provides insight into the child's overall bone health trajectory; to this end, BMD is most useful when implemented as a longitudinal monitoring tool, much like linear growth is tracked children as a barometer of overall well-being. The exact timing and frequency of screening BMD depends on the risk of secondary osteoporosis and, consequently, differs between patient groups and the individual patient. In general, the following guidelines can be followed:

1. Patients that will be treated with daily oral or IV glucocorticoids for more than three months should be considered for a baseline spine radiograph or DXA-based vertebral fracture assessment. This is recommended since the earliest reported vertebral fracture in children treated with glucocorticoids is at 4 months after the start of glucocorticoids [42]. These children should also undergo a follow-up radiograph at 12 months if risk factors are persistent since this is the time point with the highest incidence of vertebral fractures [42,43]. Annual to biennial imaging for VF is advised afterward for patients still treated with glucocorticoids; VF imaging beyond this critical high-risk period is then customized to the magnitude of risk factors thereafter. For further information on the follow-up frequency, see Ward et al. [39];
2. The same principles apply to children with other risk factors for bone fragility. In children with immobilization, especially in combination with drugs that can cause secondary osteoporosis, screening should start at the latest by 6–8 years of age and then at intervals of about two years thereafter until the end of growth or earlier in case of suffering from back pain or fragility fractures [27,44]. Monitoring is recommended to start by this time to plan to start with treatment. Treatment should be initiated before there is not enough residual growth potential to reshape vertebral bodies following VFs. Since BMD is useful as a longitudinal measurement to assist the clinician in understanding the child's overall bone health trajectory and in making logical decisions about the need for ongoing monitoring versus discharge from bone health care or intervention, it is recommended that a BMD measurement is carried out at least as frequently as spine radiographs according to the guidelines above, with assessments every 6 months in those children at greatest risk [6,45]. If the spine BMD Z-score declines by more than 0.5 on successive measurements, or there is back pain, earlier spine imaging is recommended every 1–2 years in those with persistent risk factors (as demonstrated in recent guidelines put forward in a prototypical osteoporotic condition of childhood, glucocorticoid-treated DMD) [11].

5. Prevention of Secondary Osteoporosis

Numerous interventions may prevent or reduce bone fragility fractures [46]. This starts with recognizing the risk of bone fragility and osteoporosis, as discussed above. Children with a condition or risk factors known to be associated with low bone strength or secondary osteoporosis, such as wheelchair dependence, cytokine-release related disorders, or who are on certain drugs, have an increased risk of fragility fractures [23]. General osteoporosis prevention measures are effective treatment of the underlying condition and associated morbidities, optimizing nutritional state and weight-bearing activities, and the diagnosis and treatment of endocrinopathies, including delayed puberty, growth hormone deficiency, and thyroid disorders. For children with chronic illnesses, adequate treatment of the illness is a sine qua non for osteoporosis prevention and treatment [39]. However, sometimes the disease cannot be causally treated (i.e., severe neurodevelopmental disorder), or the treatment induces bone loss and osteoporosis, such as glucocorticoids or chemotherapy. For this reason, whenever possible, treatments should be as steroid-sparing as possible. Nutritional state is an important factor. For example, a poor nutritional state (underweight or feeding difficulties) is associated with a lower BMD in children with cerebral palsy and chronic pancreatitis [26,47,48]. Other well-known risk factors for reduced bone strength are vitamin D deficiency and a shortage of dietary intake of calcium. Even in healthy children in Europe, the prevalence of vitamin deficiency is about one in every 4–5 children [49]. Children with chronic illnesses are at high risk due to a combination of limited sun exposure and often feeding problems. The recommended daily intake of vitamin D is 400–800 IU/day, depending on the 25-hydroxyvitamin D level. The optimal serum 25-hydroxyvitamin D threshold, however, remains controversial. From a practical perspective, a 25-hydroxyvitamin D level of 50 nmol/L [20 ng/mL) or more is recommended at the end of winter [39,50]. Incidental studies report a positive result of this intervention. Bianchi et al. [51] reported a significant increase in BMD in 65% of patients with Duchenne muscular dystrophy after two years of vitamin D treatment and adjustment of dietary calcium to the recommended daily dose. Calcium is an important nutrient for adequate skeletal mineralization. The recommended dietary allowance of calcium is 700 mg/day for children between 1–3 years, 1000 mg/day between 4–8 years, and 1300 mg/day for children between 9–18 years [52]. Optimal calcium intake can be achieved by an adequate diet whenever possible [39]. The role of standard calcium supplementation in healthy children has been investigated by a meta-analysis showing only a small effect on BMD, unlikely to alter fracture risk [53]; the situation in chronically ill children may be different since risk factors may "stack" towards an increased risk of bone fragility. For this reason, optimizing conservation measures to enhance bone health, such as calcium and vitamin D intake, are standard approaches in the chronic illness setting.

Physical activities, including weight-bearing activities, have an anabolic effect on the growing skeleton. These physical activities increase bone mass in healthy children [54]. The evidence of the effect of physical or weight-bearing activities in children with chronic illnesses is still insufficient [55]. It is advised to encourage activities with a low risk of falling and bodily contact in ambulatory children with osteoporosis [39]. In non-ambulant children, modest increases in BMD have been reported following weight-bearing regimes and low amplitude, high-frequency vibration therapy [56,57]. Moreover, it was shown that exercise might improve bone strength under conditions of adequate calcium intake, showing the importance of implementing these general measures in tandem [39,58].

6. Potential for Recovery

When vulnerable children receive a diagnosis of secondary osteoporosis, there is not always the need to treat with bone-targeted therapy because of the pediatric skeleton's ability to undergo recovery in both bone mass/density and shape. Case in point, the growing skeleton has the potency to reconstitute normal heights of vertebral bodies following a vertebral collapse, a phenomenon known as "vertebral body reshaping" [34]. This has been illustrated in children with acute lymphoblastic leukemia and children with inflammatory bowel disease [8,23]. For example, many children with acute lymphoblastic leukemia will undergo vertebral body reshaping following vertebral fractures because most are diagnosed at a young age (and have significant residual growth potential), and the disease (and its treatment) are usually transient. In a series of children with acute lymphoblastic leukemia, the cumulative VF incidence over six years was 32.5% [8], and complete vertebral body reshaping occurred in 77.3% of these children. Notably, the children in which the reshaping was incomplete were older (and had less residual growth potential) and had more severe degrees of vertebral collapse.

Pediatric patients with inflammatory bowel disease, in general, are older at diagnosis, and the disease is chronic with exacerbations. Vertebral fractures have been reported at diagnosis [59]. The direct effects of chronic inflammation, use of glucocorticoids, delayed puberty, and poor nutrition are contributing factors causing secondary osteoporosis. Optimizing disease control can help reshape vertebral fractures and specific bones and may be the only therapeutic intervention needed for the effective reshaping of vertebral bodies [59].

At the opposite end of the spectrum are children with chronic diseases and persistent risk factors for osteoporosis (i.e., due to presently incurable conditions), with the need for continuous treatment with glucocorticoids, for example, boys with Duchenne muscular dystrophy. In these children, there is no capacity for spontaneous vertebral body reshaping. For this reason, contemporary care includes monitoring for signs of osteoporosis at the time of initiation of glucocorticoid treatment and starting osteoporosis treatment at the first sign of a low-trauma long-bone fracture or VF [7,11,32,60]. In such children, strategies are currently being considered for the prevention of first-ever fractures.

These three different clinical scenarios show the importance of assessing whether the vulnerable child with a fragility fracture needs osteoporosis therapy. It is important to recognize that younger age, transient risk factors, normal growth and puberty, and less severe vertebral collapse are key factors for recovery without needing bone-targeted intervention (i.e., bisphosphonates).

7. Therapy

Currently, the most widely used agents for treating secondary osteoporosis in children are bisphosphonates [39]. Bisphosphonates are synthetic analogs of pyrophosphates. Classically, treatment with intravenous bisphosphonates should be considered in children with a formal diagnosis of secondary osteoporosis that manifests as at least one low-trauma vertebral or long bone fracture and limited potential for spontaneous (i.e., medication-unassisted) recovery due to older age and/or persistence of osteoporosis risk factors [6,60]. In previous years, the ISCD criteria for otherwise healthy children were erroneously applied to some children with risk factors for osteoporosis, which meant that bisphosphonate therapy was withheld from, for example, boys with DMD who had a single femur fracture (instead of two or more long-bone fractures by 10 years of age or three or more long bone fractures by 19 years of age plus a low BMD) [61,62]. However, it is now understood that these criteria were intended for otherwise healthy children so as not to over-diagnose osteoporosis in this population, and even a single low-trauma long bone fracture can represent an osteoporotic event in children with risk factors for osteoporosis [39]. Low-trauma long bone fractures and symptomatic vertebral fractures (or asymptomatic VF in high-risk settings such as CP, DMD, and glucocorticoid-treated disorders) are the most frequent indications for treatment with intravenous bisphosphonates. The primary function of bisphosphonates is to inactivate osteoclasts, resulting in cortical and trabecular bone thickening. This makes bones wider, denser, and stronger. There has been much discussion in the literature on the use of oral versus intravenous therapy with bisphosphonates. At this time, the scientific data support the use of the (more potent) intravenous bisphosphonates over oral bisphosphonates. This article is not the place for a thorough discussion on this theme, but we refer those wanting to know more to Ward et al. [39].

There are different regimens to prescribe the most-used intravenous bisphosphonate, pamidronate. The most frequently used regimen is 1 mg/kg/day for three days every four months (in total, 9 mg/kg/year). However, a regimen of 1 mg/kg every three months (in total, 4 mg/kg/year) in children with primary osteoporosis has shown a comparable effect on BMD and reduction of fragility fractures [63]. Other studies also showed a comparable effect [64,65]. A different intravenous bisphosphonate, zoledronic acid, is also used in children. Zoledronic acid can be given every six months and is preferred by some because of its greater convenience with respect to the shorter duration of the infusion and longer duration of action compared with pamidronate. Saraff et al. showed that zoledronic acid has a comparable efficacy compared to pamidronate in children with osteogenesis imperfecta [66]. Additionally, Nasomyont et al. found no difference in effect on BMD Z scores between Pamidronate and zoledronic acid in patients with primary, secondary, and glucocorticoid-induced osteoporosis [61]. Furthermore, the costs of zoledronic acid are lower [66]. Conversely, Pamidronate probably has fewer adverse drug reactions, predominantly hypocalcemia [61], nausea, and vomiting. It is important to note that the first-infusion reactions of intravenous bisphosphonate therapy can precipitate adrenal insufficiency in glucocorticoid-treated children. For this reason, steroid stress dosing is recommended in children at risk for adrenal suppression in this context. In a recent study of intravenous zoledronic acid versus intravenous placebo in glucocorticoid-treated children, 25% of children in the placebo group had at least one adverse event following the first zoledronic acid infusion [67]. This observation provided evidence that not all health events following the first intravenous bisphosphonate infusions are bisphosphonate-related but may be linked to underlying chronic illnesses.

In practice, there are still two other questions regarding the treatment with bisphosphonates in children with permanent risk factors: when should we start bisphosphonates, and how long should patients be on bisphosphonates therapy?

When should we start bisphosphonates? As said before, treatment with intravenous bisphosphonates should be considered in children fulfilling the ISCD criteria. However, in children with significant risk factors for bone fragility due to chronic illness, even a single low-trauma long bone or vertebral fracture (in the absence of a "low BMD") may be sufficient to diagnose the child with osteoporosis, as discussed in a recent paper by Ward et al. [7]. recently Nasomyont et al. published a study of the treatment with bisphosphonates in 48 with primary osteoporosis, 46 with secondary osteoporosis without glucocorticoids and 29 children with glucocorticoid-induced osteoporosis [56]. They treated 19% of the patients with intravenous bisphosphonates without these patients fulfilling the ISCD criteria for osteoporosis. One of these indications was low, declining BMD without a history of fractures; another was low BMD in association with kidney stones in children with ongoing risk factors. In addition, Draaisma et al. treated children on a ketogenic diet with ongoing risk factors with bisphosphonates out of the ISCD criteria [68].

How long should patients be on bisphosphonate therapy? Some observations in children with both primary and secondary osteoporosis have provided answers to these questions [69]. After discontinuing bisphosphonate therapy in children with open growth plates and persistent risk factors for osteoporosis, the newly formed bone has a low density. This may cause a point of increased bone fragility because of the "density differential" between the high-density bone (formed under treatment with bisphosphonates) and the low-density bone [70], leading to fractures in the transition zone between treated and untreated bone [71,72]. So, if bisphosphonate therapy should be continued, for how long? A possible answer followed from a study in children with osteogenesis imperfecta [71]. In children with still open growth plates and discontinuing bisphosphonate therapy, the BMD z-scores declined. However, discontinuing therapy in children with closed growth plates had no effect on the BMD z-scores two years later. This has led to the recommendation of continuation of bisphosphonate therapy until the closure of the growth plates in those with persistent risk factors (such as congenital/genetic disorders and long-term acquired risk factors, such as chronic glucocorticoid therapy). Starting with a higher dose until the patient is clinically stable and tapering the dose or frequency is often implemented in order to avoid over-treatment in the context of long-term therapy considered [39,73].

In summary, treatment with intravenous bisphosphonates should be considered in children fulfilling the ISCD criteria (intended for otherwise healthy children with clinically significant fractures and low, appropriately size-adjusted BMD parameters). However, it should also be considered in high-risk patients outside of this conservative approach who have even a single-low-trauma long bone or vertebral fracture, a known increased risk for bone fragility, and limited potential for spontaneous recovery [39,68]. It is important to keep in mind that not all children with secondary osteoporosis require bisphosphonate therapy following a fragility fracture; those with transient risk factors for vertebral or long bone fractures and potential for spontaneous recovery due to younger age or less severe vertebral collapse typically do not need osteoporosis intervention, unless they have significant bone pain interfering with the quality of life.

A proposition for screening, prevention, and therapy of secondary osteoporosis is given in Figure 1.

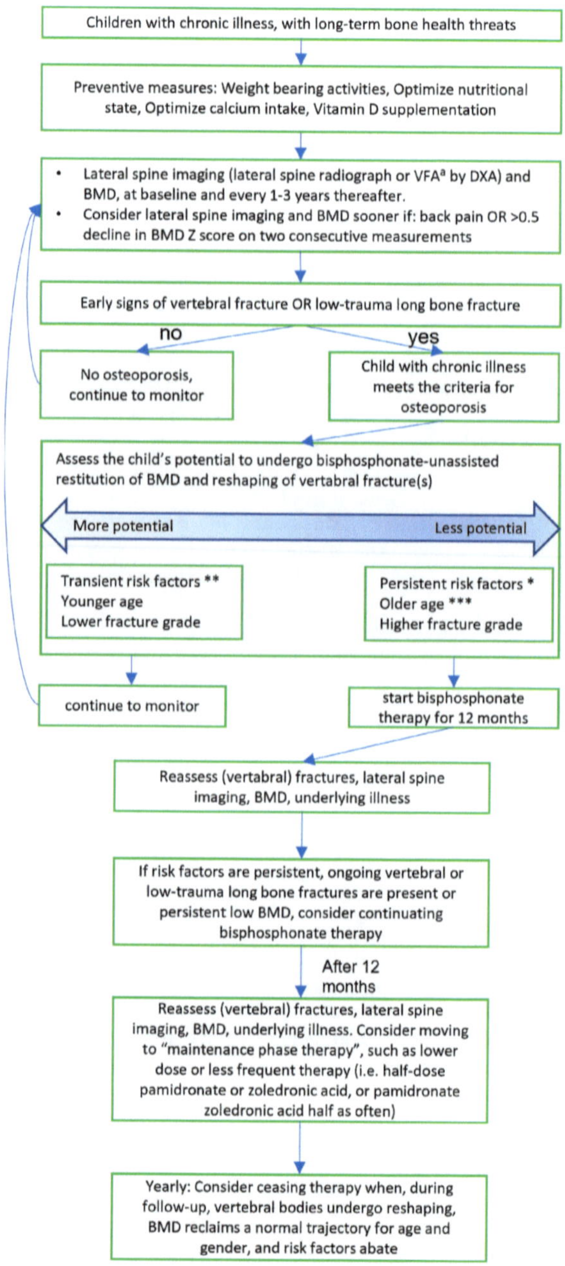

Figure 1. Recommendation for investigation and treatment of secondary osteoporosis in children. Partly based on the previous recommendation of papers of Ward [32,60], Fehlings et al. [50], and Simm et al. [73].[a] VFA = Vertebral Fracture Assessment * Persistent risk factors are >3 months steroids, sub-normal mobility, poorly controlled underlying disease ** Transient risk factors are <3months steroids, short-term immobilization (<2 weeks), well-controlled underlying disease *** Older age is defined as >8 years in girls and >9 years in boys.

8. Specific Groups

While the principles with respect to at-risk children mapped out previously in this article apply to all high-risk groups, the following describes additional disease-specific nuances in the management of patients who fall into three groups: those with profound intellectual and multiple disabilities, those with neuromuscular conditions otherwise, and children receiving a therapeutic ketogenic diet for the treatment of intractable seizures.

Profound intellectual and multiple disabilities

Children with profound intellectual and multiple disabilities (PIMD) form an extremely heterogeneous group regarding cause, level of functioning, and co-morbidities [74]. A GMFCS of four or five in combination with a severe intellectual disability (IQ < 30 or developmental age < 2 years), regardless of the underlying disease, is often used as a criterion for PIMD. Low BMD and secondary osteoporosis are highly prevalent co-morbidities in PIMD. As low BMD often goes unnoticed, a (low-impact) fracture is a common presenting sign of low BMD or osteoporosis. Since children with PIMD are frequently limited in their communication skills, there can be a significant delay in noticing these complications. Presedo et al. reported a delay in diagnosis of a fracture in half of the patients with cerebral palsy with an average delay of 10 days [75], stressing the need for a high index of suspicion in this population.

Incidence rates for fractures in children with PIMD are scarce. However, ample literature regarding fractures and low BMD in children and adults with cerebral palsy is available. As the level of (motor) functioning in severe cerebral palsy (GMFCS 4 or higher) and co-morbidities (e.g., epilepsy) in cerebral palsy in PIMD are fairly similar, these groups can be considered interchangeably regarding osteoporosis and low BMD. The incidence rates for fractures in children with severe cerebral palsy have been reported to be around 4% [76]. However, these numbers vary greatly. A systematic review by Mergler, including 21 studies in children with severe cerebral palsy, reported incidence rates of fractures varying from 2.7–23% [27]. The prevalence of low BMD was higher, with reported rates of 27–77%. Several studies have evaluated low BMD and secondary osteoporosis risk factors in children with cerebral palsy/PIMD. Although the reported significance of the risk factors varies, non-ambulatory status, feeding difficulties, and anticonvulsant use are repeatedly outlined as the foremost. Several studies have also shown tube feeding as an independent risk factor for low BMD [76,77], although it is unclear if this is a consequence of more excessive feeding difficulties. Other risk factors that have been reported are dietary deficits (calcium, vitamin D) and the use of other drugs, such as proton pump inhibitors [78], solely progestin-containing contraceptives [79], and delayed puberty. Of course, sustaining previous fractures has also been shown to be an important risk factor [80], as is low BMD.

Conversely, laboratory studies (e.g., calcium, phosphorus, alkaline phosphatase) have not been shown to correlate with low BMD. As calcium and phosphatase levels in the blood are maintained at a constant level at the expense of bone composition, this is not unimaginable, but unremarkable blood tests can give the clinician a false sense of security.

Diagnosing and following up on low BMD in children with PIMD deserves special attention. Measurement by DXA, including DXA-based vertebral fracture assessment (or baseline spine radiograph), is considered the gold standard, but its availability in hospitals is limited. Furthermore, movement during measurement, scoliosis, or contractures can negatively influence the measurements. Alternatives, such as the aforementioned digital X-ray radiogrammetry, have been proven to correlate well with DXA [81] in healthy children. This technique has also been investigated in children with PIMD and seems feasible, with the exception of children with (severe) anatomical deformities [20]. Screening for low BMD can be performed as stated before (in the section on screening).

In PIMD, it is almost without exception impossible to adequately treat the underlying disease. Following this, it is paramount to optimally treat all of the independent factors associated with low BMD. Regarding vitamin D and calcium, the recommendations mentioned in the Section 5 are valid. To adequately treat feeding difficulties, we recommend the early involvement of a multidisciplinary team, including a pediatric dietician and speech therapist.

All anticonvulsants have been, either clinically or preclinically, associated with disturbances in bone metabolism. However, enzyme-inducing anticonvulsants such as phenytoin and carbamazepine have been shown to carry a higher risk, possibly due to inducing vitamin D metabolism [82]. Multiple and prolonged anticonvulsant use (>2 years) is also associated with a higher fracture risk [83]. As in antibiotics, restraint anticonvulsant use should be considered to limit the risk of fragility fractures/low BMD. As gastro-esophageal reflux tends to diminish over time, decreasing or discontinuing proton pump inhibitors could also contribute. Lastly, systemic progestin-only contraceptives should be replaced with alternatives also containing estrogen.

Considering immobilization, loading exercises conducted by a physical therapist experienced in children with PIMD are important. Prevention of contractures should also deserve attention to maintain mobility and facilitate imaging options.

Aids, such as a standing frame implemented by a pediatric rehabilitation specialist, have also been shown to slightly improve BMD [84]. The Section 7 mentions osteoporosis treatment with bisphosphonate therapy.

Neuromuscular disorders

Children with neuromuscular disorders, including congenital myopathies and (congenital) muscular dystrophies, are at particular risk for reduced bone quality and, consequently, fragility fractures due to a combination of factors. Impaired mobility and altered musculoskeletal interaction reduce cortical thickness and prevent the preservation of the bone structure. Additionally, more myokines are expressed in atrophied muscles, leading to osteoclast differentiation and bone loss induction [85]. Further, children with a neuromuscular disease frequently experience nutritional issues due to swallowing problems and have decreased sun exposure, leading to a deficiency of calcium and vitamin D [86–89]. Children with Duchenne muscular dystrophy are particularly at risk for decreased bone strength when treated with glucocorticoids [90], as mentioned before. Next, the number of falls increases, and the falling mechanism is altered in patients with neuromuscular diseases. Ambulant patients commonly report problems with walking, poor balance, frequent trips, and falls [91,92]. Patients with neuromuscular diseases tend to have a different falling mechanism: they fall on their proximal bones because they are not fast enough to reach out their hands in order to break their fall [93]. This results in more frequent long-bone fractures of the long bone that are more proximally located (i.e., humerus, femur).

It is essential to prevent long bone fractures in children with neuromuscular diseases as far as possible in order to optimize functional prognosis, quality of life, and survival [94–96]. There are clues that limb immobilization in patients with neuromuscular disease may cause more loss of muscle mass than in healthy subjects [97]. Fractures of the lower extremities might further lead to permanent loss of ambulance [96,98]. Low bone strength can impair neuromuscular management and rehabilitation, including surgical treatments for scoliosis or foot deformities [99,100]. Pain caused by long bone fractures negatively influences rehabilitative care and, consequently, survival [94,101].

Routine assessment of bone quality in children with neuromuscular diseases is recommended. We propose a DXA-scan (including DXA-based vertebral fracture assessment where available) in all patients at diagnosis of the neuromuscular disease and subsequently at every one to two years, depending on the overall BMD trajectory (and more often in the patient is on glucocorticoid therapy as in boys with DMD). A digital X-ray radiogrammetry of the hand to measure the BHI is a feasible alternative when it is not possible to perform a DXA scan (movement during measurement, metallic implants, contractures, scoliosis) or age < 3 years (no normative values for DXA available for this age group). In order to detect vertebral compression fractures, a lateral thoracolumbar spine x-ray should be performed when DXA-based vertebral assessment is not possible.

Children using Ketogenic Diet

Treatment with a ketogenic diet (KDT) is a nonpharmacologic intervention for intractable childhood epilepsy. The diet consists of a high fat (mostly 80% fat), adequate protein, and a low-carbohydrate percentage that mimics a state of fasting. In intractable childhood epilepsy, treatment with a ketogenic diet can lead to a beneficial effect on seizure control, regardless of age or seizure type [102]. The ketogenic diet is associated with skeletal demineralization with a reduction of BMD between 0.16–0.6 z-score/year [68,103,104]. This results in an increased incidence of bone fractures in up to 21% of the children during a follow-up period of 6 years or more [105].

The ketogenic diet results in a high "acid load" via the ketone bodies, alterations in vitamin D levels, and lowering growth factors [105]. This also causes failure to accrue bone at a normal rate and increased bone loss due to increased bone absorption [106].

Prophylactic supplementation with calcium and vitamin D was considered mandatory for all children on a ketogenic diet, but this may be insufficient in preventing ongoing declines in BMD Z-scores [68,103].

Moreover, during long-term follow-up of KDT, an increased incidence of kidney stones has also been found, mostly due to hypercalciuria. Even frank hypercalcemia has been described along this spectrum [106]. Therefore, prophylactic supplementation of calcium and vitamin D is challenging. Screening with DXA scans to evaluate for a diminished BMD is recommended in children on KDT for over 2 years by 12 of the 25 [48%) centers participating in the International Ketogenic Diet Study Group [102]. The main recommendation of the recent report of this study group was that screening with DXA scans should be standard. Interventions should be considered in case of abnormal results. DXA scans should be repeated a year afterward [102].

These findings support the need to monitor bone strength in individuals on KDT, by following parameters, such as calcium intake, vitamin D deficiency, and activity levels. Developing kidney stones remains a major complication of KDT and has to be taken into account when assessing a patient on the diet [68].

9. Conclusions

Chronically ill and/or disabled children suffer from direct and indirect effects of their disease, such as immobilization, osteotoxic drug use, chronic inflammation, reduced time outdoors, and poor nutrition. All these factors may lead to bone fragility (low-trauma fractures) due to secondary osteoporosis. Early screening and prevention of declines in bone mass and BMD, plus timely identification of vertebral and long bone fractures, are the cornerstone of osteoporosis management in this context. When conservative preventive measures fail, treatment with intravenous bisphosphonates should be considered in at-risk children with the diagnosis of secondary osteoporosis (Figure 1), where even a single long bone or vertebral fracture provides a sufficient rationale to initiate treatment in children when there are significant risk factors for bone fragility plus limited potential for spontaneous recovery. Looking ahead, prevention of first-ever fractures is on the minds of clinicians caring for the highest-risk patients, such as boys with Duchenne muscular dystrophy, either through relatively steroid-sparing therapy [107] or via pre-fracture initiation of bisphosphonate therapy. Studies have facilitated the sightline to the prevention of first fractures in high-risk children, which have highlighted the children who are most likely to recover from osteoporosis, obviating the need for osteoporosis therapy versus those with the least potential for recovery, underscoring the need for anticipatory prevention.

Author Contributions: Conceptualization: A.T.M.D. and J.M.T.D.; methodology: A.T.M.D. and J.M.T.D.; formal and statistical analysis: not performed; investigation: A.T.M.D., E.J.M.J., J.G., K.B. and J.M.T.D.; data curation: not performed; writing—original draft preparation: A.T.M.D., E.J.M.J., J.G., K.B. and J.M.T.D.; writing—review and editing: A.T.M.D., E.J.M.J., J.G., K.B., L.M.W. and J.M.T.D.; supervision: A.T.M.D. and J.M.T.D. All authors have read and agreed to the published version of the manuscript.

Funding: This research received no external funding.

Institutional Review Board Statement: An Institutional Review Board Statement was not necessary due to the type of article.

Informed Consent Statement: Informed consent was obtained from all subjects involved in the study.

Data Availability Statement: The data presented in this study are openly available from the author.

Acknowledgments: Jos Draaisma and Joyce Geelen are members of the European Reference Network for Developmental Anomalies and Intellectual Disability (ERN-ITHACA).

Conflicts of Interest: All the authors completed the ICMJE uniform disclosure form. The authors have no conflict of interest to declare. In the last two years, Dr. Ward has been a consultant to and/or participated in clinical trials with Amgen, PTC, ReveraGen, and Santhera.

References

1. Vulnerable populations: Who are they? *Am. J. Manag.Care* **2006**, *12* (Suppl. S13), S348–S352.
2. Barnett, K.; Mercer, S.W.; Norbury, M.; Watt, G.; Wyke, S.; Guthrie, B. Epidemiology of multimorbidity and implications for health care, research, and medical education: A cross-sectional study. *Lancet* **2012**, *380*, 37–43. [CrossRef] [PubMed]
3. Bell, A.; Kendler, D.L.; Khan, A.A.; Shapiro, M.C.M.; Morisset, A.; Leung, J.-P.; Reiner, M.; Colgan, S.M.; Slatkovska, L.; Packalen, M. A retrospective observational study of osteoporosis management after a fragility fracture in primary care. *Arch. Osteoporos.* **2022**, *17*, 75. [CrossRef] [PubMed]
4. Sochett, E.B.; Mäkitie, O. Osteoporosis in chronically ill children. *Ann. Med.* **2005**, *37*, 286–294. [CrossRef]
5. Swain, M.; Kamper, S.J.; Maher, C.G.; Broderick, C.; McKay, D.; Henschke, N. Relationship between growth, maturation and musculoskeletal conditions in adolescents: A systematic review. *Br. J. Sport. Med.* **2018**, *52*, 1246–1252. [CrossRef]
6. Bishop, N.; Arundel, P.; Clark, E.; Dimitri, P.; Farr, J.; Jones, G.; Makitie, O.; Munns, C.F.; Shaw, N. Fracture prediction and the definition of osteoporosis in children and adolescents: The ISCD 2013 Pediatric Official Positions. *J. Clin. Densitom.* **2014**, *17*, 275–280. [CrossRef]
7. Ward, L.M.; Weber, D.R.; Munns, C.F.; Högler, W.; Zemel, B.S. A Contemporary View of the Definition and Diagnosis of Osteoporosis in Children and Adolescents. *J. Clin. Endocrinol. Metab.* **2020**, *105*, e2088–e2097. [CrossRef]
8. Ward, L.M.; Ma, J.; Lang, B.; Ho, J.; Alos, N.; Matzinger, M.A.; Shenouda, N.; Lentle, B.; Jaremko, J.L.; Wilson, B. Bone Morbidity and Recovery in Children with Acute Lymphoblastic Leukemia: Results of a Six-Year Prospective Cohort Study. *J. Bone Miner. Res.* **2018**, *33*, 1435–1443. [CrossRef]
9. Genant, H.K.; Wu, C.Y.; van Kuijk, C.; Nevitt, M.C. Vertebral fracture assessment using a semiquantitative technique. *J. Bone Miner. Res.* **1993**, *8*, 1137–1148. [CrossRef]
10. Ward, L.M.; Ma, J.; Robinson, M.-E.; Scharke, M.; Ho, J.; Houghton, K.; Huber, A.; Scuccimarri, R.; Barsalou, J.; Roth, J. Osteoporotic Fractures and Vertebral Body Reshaping in Children with Glucocorticoid-treated Rheumatic Disorders. *J. Clin. Endocrinol. Metab.* **2021**, *106*, e5195–e5207. [CrossRef]
11. Birnkrant, D.J.; Bushby, K.; Bann, C.M.; Alman, B.A.; Apkon, S.D.; Blackwell, A.; Case, L.E.; Cripe, L.; Hadiyannakis, S.; Olson, A.K. Diagnosis and management of Duchenne muscular dystrophy, part 2: Respiratory, cardiac, bone health, and orthopaedic management. *Lancet Neurol.* **2018**, *17*, 347–361. [CrossRef] [PubMed]
12. Crabtree, N.; Chapman, S.; Högler, W.; Hodgson, K.; Chapman, D.; Bebbington, N.; Shaw, N.J. Vertebral fractures assessment in children: Evaluation of DXA imaging versus conventional spine radiography. *Bone* **2017**, *97*, 168–174. [CrossRef] [PubMed]
13. Weber, D.R.; Boyce, A.; Gordon, C.; Högler, W.; Kecskemethy, H.H.; Misra, M.; Swolin-Eide, D.; Tebben, P.; Ward, L.M.; Wasserman, H. The Utility of DXA Assessment at the Forearm, Proximal Femur, and Lateral Distal Femur, and Vertebral Fracture Assessment in the Pediatric Population: 2019 ISCD Official Position. *J. Clin. Densitom.* **2019**, *22*, 567–589. [CrossRef] [PubMed]
14. van Rijn, R.R.; Van Kuijk, C. Of small bones and big mistakes; bone densitometry in children revisited. *Eur. J. Radiol.* **2009**, *71*, 432–439. [CrossRef] [PubMed]
15. Henderson, R.C.; Berglund, L.M.; May, R.; Zemel, B.S.; I Grossberg, R.; Johnson, J.; Plotkin, H.; Stevenson, R.D.; Szalay, E.; Wong, B. The relationship between fractures and DXA measures of BMD in the distal femur of children and adolescents with cerebral palsy or muscular dystrophy. *J. Bone Miner. Res.* **2010**, *25*, 520–526. [CrossRef]
16. Messina, C.; Lastella, G.; Sorce, S.; Piodi, L.P.; Rodari, G.; Giavoli, C.; Marchelli, D.; Guglielmi, G.; Ulivieri, F.M. Pediatric dual-energy X-ray absorptiometry in clinical practice: What the clinicians need to know. *Eur. J. Radiol.* **2018**, *105*, 153–161. [CrossRef]
17. Schündeln, M.M.; Marschke, L.; Bauer, J.J.; Hauffa, P.K.; Schweiger, B.; Führer-Sakel, D.; Lahner, H.; Poeppel, T.D.; Kiewert, C.; Hauffa, B.P. A Piece of the Puzzle: The Bone Health Index of the BoneXpert Software Reflects Cortical Bone Mineral Density in Pediatric and Adolescent Patients. *PLoS ONE* **2016**, *11*, e0151936. [CrossRef]
18. Leijten, A.D.; Hampsink, B.; Janssen, M.; Klein, W.M.; Draaisma, J.M.T. Can digital X-ray radiogrammetry be an alternative for dual-energy X-ray absorptiometry in the diagnosis of secondary low bone quality in children? *Eur. J. Pediatr.* **2019**, *178*, 1433–1441. [CrossRef]

19. Mergler, S.; Rieken, R.; Tibboel, D.; Evenhuis, H.M.; van Rijn, R.R.; Penning, C. Lumbar spine and total-body dual-energy X-ray absorptiometry in children with severe neurological impairment and intellectual disability: A pilot study of artefacts and disrupting factors. *Pediatr. Radiol.* **2012**, *42*, 574–583. [CrossRef]
20. Mergler, S.; de Man, S.A.; Boot, A.M.; Heus, K.G.C.B.B.-D.; Huijbers, W.A.R.; van Rijn, R.R.; Penning, C.; Evenhuis, H.M. Automated radiogrammetry is a feasible method for measuring bone quality and bone maturation in severely disabled children. *Pediatr. Radiol.* **2016**, *46*, 1017–1022. [CrossRef]
21. Bachrach, L.K.; Gordon, C.M. Bone Densitometry in Children and Adolescents. *Pediatrics* **2016**, *138*, e20162398. [CrossRef] [PubMed]
22. Shalof, H.; Dimitri, P.; Shuweihdi, F.; Offiah, A.C. Which skeletal imaging modality is best for assessing bone health in children and young adults compared to DXA? A systematic review and meta-analysis. *Bone* **2021**, *150*, 116013. [CrossRef] [PubMed]
23. Digby, M.G.; Bishop, N.J.; Paggiosi, M.A.; Offiah, A.C. HR-pQCT: A non-invasive 'biopsy' to assess bone structure and strength. *Arch. Dis. Child. Educ. Pract. Ed.* **2016**, *101*, 268–270. [CrossRef] [PubMed]
24. Palisano, R.; Rosenbaum, P.; Walter, S.; Russell, D.; Wood, E.; Galuppi, B. Development and reliability of a system to classify gross motor function in children with cerebral palsy. *Dev. Med. Child Neurol.* **1997**, *39*, 214–223. [CrossRef]
25. Saraff, V.; Högler, W. Endocrinology and Adolescence: Osteoporosis in children: Diagnosis and management. *Eur. J. Endocrinol.* **2015**, *173*, R185–R197. [CrossRef]
26. Henderson, R.C.; Lark, R.K.; Gurka, M.J.; Worley, G.; Fung, E.B.; Conaway, M.; Stallings, V.A.; Stevenson, R.D. Bone density and metabolism in children and adolescents with moderate to severe cerebral palsy. *Pediatrics* **2002**, *110*, e5. [CrossRef]
27. Mergler, S.; Evenhuis, H.M.; Boot, A.M.; De Man, S.A.; Heus, K.G.B.-D.; Huijbers, W.A.; Penning, C. Epidemiology of low bone mineral density and fractures in children with severe cerebral palsy: A systematic review. *Dev. Med. Child Neurol.* **2009**, *51*, 773–778. [CrossRef]
28. Larson, C.M.; Henderson, R.C. Bone Mineral Density and Fractures in Boys with Duchenne Muscular Dystrophy. *J. Pediatr. Orthop.* **2000**, *20*, 71–74. [CrossRef]
29. Crabtree, N.J.; Roper, H.; Shaw, N.J. Cessation of ambulation results in a dramatic loss of trabecular bone density in boys with Duchenne muscular dystrophy (DMD). *Bone* **2022**, *154*, 116248. [CrossRef]
30. Joseph, S.; Wang, C.; Bushby, K.; Guglieri, M.; Horrocks, I.; Straub, V.; Ahmed, S.F.; Wong, S.C. Fractures and Linear Growth in a Nationwide Cohort of Boys with Duchenne Muscular Dystrophy with and Without Glucocorticoid Treatment: Results from the UK NorthStar Database. *JAMA Neurol.* **2019**, *76*, 701–709. [CrossRef]
31. Brent, M.B.; Brüel, A.; Thomsen, J.S. A Systematic Review of Animal Models of Disuse-Induced Bone Loss. *Calcif. Tissue Int.* **2021**, *108*, 561–575. [CrossRef] [PubMed]
32. Ward, L.M. Part I: Which Child with a Chronic Disease Needs Bone Health Monitoring? *Curr. Osteoporos. Rep.* **2021**, *19*, 278–288. [CrossRef] [PubMed]
33. Lorenzo, J. Cytokines and Bone: Osteoimmunology. *Handb. Exp. Pharmacol.* **2020**, *262*, 177–230. [PubMed]
34. Ward, L.M. Glucocorticoid-Induced Osteoporosis: Why Kids Are Different. *Front. Endocrinol.* **2020**, *11*, 576. [CrossRef]
35. Compston, J. Glucocorticoid-induced osteoporosis: An update. *Endocrine* **2018**, *61*, 7–16. [CrossRef]
36. van Staa, T.P.; Leufkens, H.G.; Cooper, C. The epidemiology of corticosteroid-induced osteoporosis: A meta-analysis. *Osteoporos. Int.* **2002**, *13*, 777–787. [CrossRef]
37. Weber, F.J.; Latshang, T.D.; Blum, M.R.; Kohler, M.; Wertli, M.M. Prognostic factors, disease course, and treatment efficacy in Duchenne muscular dystrophy: A systematic review and meta-analysis. *Muscle Nerve* **2022**, *66*, 462–470. [CrossRef]
38. Singh, A.; Schaeffer, E.K.; Reilly, C.W. Vertebral Fractures in Duchenne Muscular Dystrophy Patients Managed With Deflazacort. *J. Pediatr. Orthop.* **2018**, *38*, 320–324. [CrossRef]
39. Ward, L.M.; Konji, V.N.; Ma, J. The management of osteoporosis in children. *Osteoporos. Int.* **2016**, *27*, 2147–2179. [CrossRef]
40. Wood, C.L.; Soucek, O.; Wong, S.C.; Zaman, F.; Farquharson, C.; Savendahl, L.; Ahmed, S.F. Animal models to explore the effects of glucocorticoids on skeletal growth and structure. *J. Endocrinol.* **2018**, *236*, R69–R91. [CrossRef]
41. Ma, J.; Siminoski, K.; Alos, N.; Halton, J.; Ho, J.; Lentle, B.; Matzinger, M.A.; Shenouda, N.; Atkinson, S.; Barr, R. The choice of normative pediatric reference database changes spine bone mineral density Z-scores but not the relationship between bone mineral density and prevalent vertebral fractures. *J. Clin. Endocrinol. Metab.* **2015**, *100*, 1018–1027. [CrossRef]
42. Rodd, C.; Lang, B.; Ramsay, T.; Alos, N.; Huber, A.M.; Cabral, D.A.; Scuccimarri, R.; Miettunen, P.M.; Roth, J.; Atkinson, S.A. Incident vertebral fractures among children with rheumatic disorders 12 months after glucocorticoid initiation: A national observational study. *Arthritis Care Res.* **2012**, *64*, 122–131. [CrossRef] [PubMed]
43. Cummings, E.; Ma, J.; Fernandez, C.V.; Halton, J.; Alos, N.; Miettunen, P.M.; Jaremko, J.L.; Ho, J.; Shenouda, N.; Matzinger, M.A. Incident Vertebral Fractures in Children with Leukemia During the Four Years Following Diagnosis. *J. Clin. Endocrinol. Metab.* **2015**, *100*, 3408–3417. [CrossRef] [PubMed]
44. Berkvens, J.J.L.; Mergler, S.; Beerhorst, K.; Verschuure, P.; Tan, I.Y.; Majoie, H.J.M.; van den Bergh, J.P.W. Bone mineral density and fractures in institutionalised children with epilepsy and intellectual disability. *J. Intellect. Disabil. Res.* **2021**, *65*, 962–970. [CrossRef] [PubMed]
45. Bianchi, M.L.; Leonard, M.B.; Bechtold, S.; Högler, W.; Mughal, M.Z.; Schönau, E.; Sylvester, F.A.; Vogiatzi, M.; van den Heuvel-Eibrink, M.M.; Ward, L. Bone health in children and adolescents with chronic diseases that may affect the skeleton: The 2013 ISCD Pediatric Official Positions. *J. Clin. Densitom.* **2014**, *17*, 281–294. [CrossRef]

46. Ciancia, S.; Högler, W.; Sakkers, R.J.B.; Appelman-Dijkstra, N.M.; Boot, A.M.; Sas, T.C.J.; Renes, J.S. Osteoporosis in children and adolescents: How to treat and monitor? *Eur. J. Pediatr.* **2022**, *182*, 501–511. [CrossRef]
47. Fernández, J.M.M.d.Z.; Arnal, I.R.; Segura, J.L.P.; Romero, R.G.; Martínez, G.R. Bone health impairment in patients with cerebral palsy. *Arch. Osteoporos.* **2020**, *15*, 91. [CrossRef]
48. Srivastava, A.; Saini, N.; Mathias, A.; Arya, A.; Jain, S.; Yachha, S.K. Prevalence and predictive factors of undernutrition and low bone mineral density in children with chronic pancreatitis. *Pancreatology* **2021**, *21*, 74–80. [CrossRef]
49. Castano, L.; Madariaga, L.; Grau, G.; García-Castaño, A. 25(OH)Vitamin D Deficiency and Calcifediol Treatment in Pediatrics. *Nutrients* **2022**, *14*, 1854. [CrossRef]
50. Fehlings, D.; Switzer, L.; Agarwal, P.; Wong, C.; Sochett, E.; Stevenson, R.; Sonnenberg, L.; Smile, S.; Young, E.; Huber, J. Informing evidence-based clinical practice guidelines for children with cerebral palsy at risk of osteoporosis: A systematic review. *Dev. Med. Child Neurol.* **2012**, *54*, 106–116. [CrossRef]
51. Bianchi, M.L.; Morandi, L.; Andreucci, E.; Vai, S.; Frasunkiewicz, J.; Cottafava, R. Low bone density and bone metabolism alterations in Duchenne muscular dystrophy: Response to calcium and vitamin D treatment. *Osteoporos. Int.* **2011**, *22*, 529–539. [CrossRef]
52. Institute of Medicine (US) Committee to Review Dietary Reference Intakes for Vitamin D; Calcium. The National Academies Collection: Reports Funded by National Institutes of Health. In *Dietary Reference Intakes for Calcium and Vitamin D*; Ross, A.C., Taylor, C.L., Yaktine, A.L., Del Valle, H.B., Eds.; National Academies Press: Washington, DC, USA, 2011.
53. Winzenberg, T.; Shaw, K.; Fryer, J.; Jones, G. Effects of calcium supplementation on bone density in healthy children: Meta-analysis of randomised controlled trials. *BMJ* **2006**, *333*, 775. [CrossRef]
54. Tan, V.P.; Macdonald, H.M.; Kim, S.; Nettlefold, L.; Gabel, L.; Ashe, M.C.; McKay, H.A. Influence of physical activity on bone strength in children and adolescents: A systematic review and narrative synthesis. *J. Bone Miner. Res.* **2014**, *29*, 2161–2181. [CrossRef]
55. Ozel, S.; Switzer, L.; Macintosh, A.; Fehlings, D. Informing evidence-based clinical practice guidelines for children with cerebral palsy at risk of osteoporosis: An update. *Dev. Med. Child Neurol.* **2016**, *58*, 918–923. [CrossRef] [PubMed]
56. Hough, J.P.; Boyd, R.N.; Keating, J.L. Systematic review of interventions for low bone mineral density in children with cerebral palsy. *Pediatrics* **2010**, *125*, e670–e678. [CrossRef] [PubMed]
57. Ward, K.; Alsop, C.; Caulton, J.; Rubin, C.; Adams, J.; Mughal, Z. Low magnitude mechanical loading is osteogenic in children with disabling conditions. *J. Bone Miner. Res.* **2004**, *19*, 360–369. [CrossRef]
58. Julián-Almárcegui, C.; Gómez-Cabello, A.; Huybrechts, I.; González-Agüero, A.; Kaufman, J.M.; Casajús, J.A.; Vicente-Rodríguez, G. Combined effects of interaction between physical activity and nutrition on bone health in children and adolescents: A systematic review. *Nutr. Rev.* **2015**, *73*, 127–139. [CrossRef] [PubMed]
59. Wong, S.C.; Catto-Smith, A.G.A.; Zacharin, M. Pathological fractures in paediatric patients with inflammatory bowel disease. *Eur. J. Pediatr.* **2014**, *173*, 141–151. [CrossRef]
60. Ward, L.M. Part 2: When Should Bisphosphonates Be Used in Children with Chronic Illness Osteoporosis? *Curr. Osteoporos. Rep.* **2021**, *19*, 289–297. [CrossRef]
61. Nasomyont, N.; Hornung, L.N.; Gordon, C.M.; Wasserman, H. Outcomes following intravenous bisphosphonate infusion in pediatric patients: A 7-year retrospective chart review. *Bone* **2019**, *121*, 60–67. [CrossRef]
62. Morris, H.F. Veterans Administration Cooperative Studies Project No. 147. Part VI: Laboratory costs of castings from noble and alternative ceramic metal alloys. *J. Prosthet. Dent.* **1988**, *60*, 164–171. [CrossRef] [PubMed]
63. Martinez-Soto, T.; Pacaud, D.; Stephure, D.; Trussell, R.; Huang, C. Treatment of symptomatic osteoporosis in children: A comparison of two pamidronate dosage regimens. *J. Pediatr. Endocrinol. Metab.* **2011**, *24*, 271–274. [CrossRef] [PubMed]
64. Chattopadhyay, A.; Bhansali, A.; Mohanty, S.K.; Khandelwal, N.; Mathur, S.K.; Dash, R.J. Hypophosphatemic rickets and osteomalacia in polyostotic fibrous dysplasia. *J. Pediatr. Endocrinol. Metab.* **2003**, *16*, 893–896. [CrossRef] [PubMed]
65. Steelman, J.; Zeitler, P. Treatment of symptomatic pediatric osteoporosis with cyclic single-day intravenous pamidronate infusions. *J. Pediatr.* **2003**, *142*, 417–423. [CrossRef] [PubMed]
66. Saraff, V.; Sahota, J.; Crabtree, N.; Sakka, S.; Shaw, N.J.; Högler, W. Efficacy and treatment costs of zoledronate versus pamidronate in paediatric osteoporosis. *Arch. Dis. Child.* **2018**, *103*, 92–94. [CrossRef] [PubMed]
67. Ward, L.M.; Choudhury, A.; Alos, N.; Cabral, D.A.; Rodd, C.; Sbrocchi, A.M.; Taback, S.; Padidela, R.; Shaw, N.J.; Hosszu, E. Zoledronic Acid vs Placebo in Pediatric Glucocorticoid-induced Osteoporosis: A Randomized, Double-blind, Phase 3 Trial. *J. Clin. Endocrinol. Metab.* **2021**, *106*, e5222–e5235. [CrossRef]
68. Draaisma, J.M.; Hampsink, B.M.; Janssen, M.; van Houdt, N.B.; Linders, E.T.; Willemsen, M.A. The Ketogenic Diet and Its Effect on Bone Mineral Density: A Retrospective Observational Cohort Study. *Neuropediatrics* **2019**, *50*, 353–358. [CrossRef]
69. Robinson, M.-E.; Trejo, P.; Palomo, T.; Glorieux, F.H.; Rauch, F. Osteogenesis Imperfecta: Skeletal Outcomes After Bisphosphonate Discontinuation at Final Height. *J. Bone Miner. Res.* **2019**, *34*, 2198–2204. [CrossRef]
70. Rauch, F.; Munns, C.; Land, C.; Glorieux, F.H. Pamidronate in children and adolescents with osteogenesis imperfecta: Effect of treatment discontinuation. *J. Clin. Endocrinol. Metab.* **2006**, *91*, 1268–1274. [CrossRef]
71. Rauch, F.; Cornibert, S.; Cheung, M.; Glorieux, F.H. Long-bone changes after pamidronate discontinuation in children and adolescents with osteogenesis imperfecta. *Bone* **2007**, *40*, 821–827. [CrossRef]

72. Harcke, H.T.; Stevenson, K.L.; Kecskemethy, H.H.; Bachrach, S.J.; Grissom, L.E. Fracture after bisphosphonate treatment in children with cerebral palsy: The role of stress risers. *Pediatr. Radiol.* **2012**, *42*, 76–81. [CrossRef]
73. Simm, P.J.; Biggin, A.; Zacharin, M.R.; Rodda, C.P.; Tham, E.; Siafarikas, A.; Jefferies, C.; Hofman, P.L.; Jensen, D.E.; Woodhead, H. Consensus guidelines on the use of bisphosphonate therapy in children and adolescents. *J. Paediatr. Child Heal.* **2018**, *54*, 223–233. [CrossRef] [PubMed]
74. Nakken, H.; Vlaskamp, C. A Need for a Taxonomy for Profound Intellectual and Multiple Disabilities. *J. Policy Pr. Intellect. Disabil.* **2007**, *4*, 83–87. [CrossRef]
75. Presedo, A.; Dabney, K.W.; Miller, F. Fractures in patients with cerebral palsy. *J. Pediatr. Orthop.* **2007**, *27*, 147–153. [CrossRef]
76. Stevenson, R.D.; Conaway, M.; Barrington, J.W.; Cuthill, S.L.; Worley, G.; Henderson, R.C. Fracture rate in children with cerebral palsy. *Pediatr. Rehabilitation* **2006**, *9*, 396–403. [CrossRef] [PubMed]
77. Sakai, T.; Honzawa, S.; Kaga, M.; Iwasaki, Y.; Masuyama, T. Osteoporosis pathology in people with severe motor and intellectual disability. *Brain Dev.* **2020**, *42*, 256–263. [CrossRef]
78. Selvin, S.T.; Thomas, S.; Bikeyeva, V.; Abdullah, A.; Radivojevic, A.; Jad, A.A.A.; Ravanavena, A.; Ravindra, C.; Igweonu-Nwakile, E.; Ali, S. Establishing the Association Between Osteoporosis and Peptic Ulcer Disease: A Systematic Review. *Cureus* **2022**, *14*, e27188.
79. Arvio, M.; Kilpinen-Loisa, P.; Tiitinen, A.; Huovinen, K.; Mäkitie, O. Bone mineral density and sex hormone status in intellectually disabled women on progestin-induced amenorrhea. *Acta Obstet. Gynecol. Scand.* **2009**, *88*, 428–433. [CrossRef]
80. Henderson, R.C.; Kairalla, J.; Abbas, A.; Stevenson, R.D. Predicting low bone density in children and young adults with quadriplegic cerebral palsy. *Dev. Med. Child Neurol.* **2004**, *46*, 416–419. [CrossRef]
81. Thodberg, H.H.; van Rijn, R.R.; Tanaka, T.; Martin, D.D.; Kreiborg, S. A paediatric bone index derived by automated radiogrammetry. *Osteoporos. Int.* **2010**, *21*, 1391–1400. [CrossRef]
82. Petty, S.J.; Paton, L.M.; O'Brien, T.; Makovey, J.; Erbas, B.; Sambrook, P.; Berkovic, S.F.; Wark, D.W. Effect of antiepileptic medication on bone mineral measures. *Neurology* **2005**, *65*, 1358–1365. [CrossRef]
83. Beerhorst, K.; Schouwenaars, F.M.; Tan, I.Y.; Aldenkamp, A.P. Epilepsy: Fractures and the role of cumulative antiepileptic drug load. *Acta Neurol. Scand.* **2012**, *125*, 54–59. [CrossRef] [PubMed]
84. Caulton, J.M.; Ward, K.; Alsop, C.W.; Dunn, G.; E Adams, J.; Mughal, M.Z. A randomised controlled trial of standing programme on bone mineral density in non-ambulant children with cerebral palsy. *Arch. Dis. Child.* **2004**, *89*, 131–135. [CrossRef] [PubMed]
85. Qin, Y.; Peng, Y.; Zhao, W.; Pan, J.; Ksiezak-Reding, H.; Cardozo, C.; Wu, Y.; Divieti Pajevic, P.; Bonewald, L.F.; Bauman, W.A. Myostatin inhibits osteoblastic differentiation by suppressing osteocyte-derived exosomal microRNA-218: A novel mechanism in muscle-bone communication. *J. Biol. Chem.* **2017**, *292*, 11021–11033. [CrossRef] [PubMed]
86. Chou, E.; Lindeback, R.; D'Silva, A.M.; Sampaio, H.; Neville, K.; Farrar, M.A. Growth and nutrition in pediatric neuromuscular disorders. *Clin. Nutr.* **2021**, *40*, 4341–4348. [CrossRef]
87. Kooi-van Es, M.; Erasmus, C.E.; de Swart, B.J.M.; Voet, N.B.M.; van der Wees, P.J.; de Groot, I.J.M.; van den Engel-Hoek, L. Dysphagia and Dysarthria in Children with Neuromuscular Diseases, a Prevalence Study. *J. Neuromuscul Dis.* **2020**, *7*, 287–295. [CrossRef]
88. Bian, Q.; McAdam, L.; Grynpas, M.; Mitchell, J.; Harrington, J. Increased Rates of Vitamin D Insufficiency in Boys With Duchenne Muscular Dystrophy Despite Higher Vitamin D(3) Supplementation. *Glob. Pediatr. Heal.* **2019**, *6*, 2333794x19835661. [CrossRef]
89. Lips, P.; van Schoor, N.M. The effect of vitamin D on bone and osteoporosis. *Best Pract. Res. Clin. Endocrinol. Metab.* **2011**, *25*, 585–591. [CrossRef]
90. Ward, L.M.; Hadjiyannakis, S.; McMillan, H.J.; Noritz, G.; Weber, D.R. Bone Health and Osteoporosis Management of the Patient with Duchenne Muscular Dystrophy. *Pediatrics* **2018**, *142* (Suppl. S2), S34–S42. [CrossRef]
91. Kennedy, R.A.; Carroll, K.; McGinley, J.L.; Paterson, K.L. Walking and weakness in children: A narrative review of gait and functional ambulation in paediatric neuromuscular disease. *J. Foot Ankle Res.* **2020**, *13*, 10. [CrossRef]
92. Wiles, C.M.; E Busse, M.; Sampson, C.M.; Rogers, M.T.; Fenton-May, J.; van Deursen, R. Falls and stumbles in myotonic dystrophy. *J. Neurol. Neurosurg. Psychiatry* **2006**, *77*, 393–396. [CrossRef] [PubMed]
93. Horlings, C.G.; Küng, U.M.; van Engelen, B.G.; Voermans, N.C.; Hengstman, G.J.; van der Kooi, A.J.; Bloed, B.R.; Allum, J.H.J. Balance control in patients with distal versus proximal muscle weakness. *Neuroscience* **2009**, *164*, 1876–1886. [CrossRef] [PubMed]
94. Wang, C.H.; Dowling, J.J.; North, K.; Schroth, M.K.; Sejersen, T.; Shapiro, F.; Bellini, J.; Weiss, H.; Guillet, M.; Amburgery, K. Consensus statement on standard of care for congenital myopathies. *J. Child Neurol.* **2012**, *27*, 363–382. [CrossRef] [PubMed]
95. Iolascon, G.; Paoletta, M.; Liguori, S.; Curci, C.; Moretti, A. Neuromuscular Diseases and Bone. *Front. Endocrinol.* **2019**, *10*, 794. [CrossRef] [PubMed]
96. Vestergaard, P.; Glerup, H.; Steffensen, B.F.; Rejnmark, L.; Rahbek, J.; Moseklide, L. Fracture risk in patients with muscular dystrophy and spinal muscular atrophy. *J. Rehabil. Med.* **2001**, *33*, 150–155. [CrossRef]
97. Joya, J.E.; Kee, A.J.; Nair-Shalliker, V.; Ghoddusi, M.; Nguyen, M.-A.T.; Luther, P.; Hardeman, E.C. Muscle weakness in a mouse model of nemaline myopathy can be reversed with exercise and reveals a novel myofiber repair mechanism. *Hum. Mol. Genet.* **2004**, *13*, 2633–2645. [CrossRef] [PubMed]
98. Hsu, J.D. Extremity fractures in children with neuromuscular disease. *Johns Hopkins Med. J.* **1979**, *145*, 89–93.
99. Vialle, R.; Thévenin-Lemoine, C.; Mary, P. Neuromuscular scoliosis. *Orthop. Traumatol. Surg. Res.* **2013**, *99* (Suppl. S1), S124–S139. [CrossRef]

100. Strømsøe, K. Fracture fixation problems in osteoporosis. *Injury* **2004**, *35*, 107–113. [CrossRef]
101. Pianucci, L.; Sonagra, M.; Greenberg, B.A.; Priestley, D.R.; Gmuca, S. Disordered eating among adolescents with chronic pain: The experience of a pediatric rheumatology subspecialty pain clinic. *Pediatr. Rheumatol. Online J.* **2021**, *19*, 16. [CrossRef]
102. Kossoff, E.H.; Zupec-Kania, B.A.; Auvin, S.; Ballaban-Gil, K.R.; Bergqvist, A.G.C.; Blackford, R.; Buchhalter, J.R.; Caraballo, R.H.; Cross, J.H.; Dahlin, M.G. Optimal clinical management of children receiving dietary therapies for epilepsy: Updated recommendations of the International Ketogenic Diet Study Group. *Epilepsia Open* **2018**, *3*, 175–192. [CrossRef]
103. Bergqvist, A.C.; Schall, J.I.; Stallings, V.A.; Zemel, B.S. Progressive bone mineral content loss in children with intractable epilepsy treated with the ketogenic diet. *Am. J. Clin. Nutr.* **2008**, *88*, 1678–1684. [CrossRef] [PubMed]
104. Simm, P.J.; Bicknell-Royle, J.; Lawrie, J.; Nation, J.; Draffin, K.; Stewart, K.G.; Cameron, F.J.; Scheffer, I.E.; Mackay, M.T. The effect of the ketogenic diet on the developing skeleton. *Epilepsy Res.* **2017**, *136*, 62–66. [CrossRef] [PubMed]
105. Groesbeck, D.K.; Bluml, R.M.; Kossoff, E.H. Long-term use of the ketogenic diet in the treatment of epilepsy. *Dev. Med. Child Neurol.* **2006**, *48*, 978–981. [CrossRef] [PubMed]
106. Hawkes, C.P.; Roy, S.M.; Dekelbab, B.; Frazier, B.; Grover, M.; Haidet, J.; Listman, J.; Maden, S.; Roan, M.; Rodd, C. Hypercalcemia in Children Using the Ketogenic Diet: A Multicenter Study. *J. Clin. Endocrinol. Metab.* **2021**, *106*, e485–e495. [CrossRef] [PubMed]
107. Guglieri, M.; Clemens, P.R.; Perlman, S.J.; Smith, E.C.; Horrocks, I.; Finkel, R.S.; Mah, J.K.; Deconink, N.; Goemans, N.; Haberlova, J. Efficacy and Safety of Vamorolone vs Placebo and Prednisone Among Boys with Duchenne Muscular Dystrophy: A Randomized Clinical Trial. *JAMA Neurol.* **2022**, *79*, 1005–1014. [CrossRef]

applied sciences

MDPI

Brief Report

Eye Health Screening in Migrant Population: Primary Care Experience in Lazio (Italy) from the PROTECT Project

Alice Bruscolini [1], Giacomo Visioli [1,*], Marco Marenco [1], Veronica Cherubini [1], Anna Maria Comberiati [1], Gaspare Palaia [2], Massimo Ralli [1], Livia Ottolenghi [2], Alessandro Lambiase [1,*] and Antonella Polimeni [2]

1 Department of Sense Organs, Sapienza—University of Rome, 00161 Rome, Italy
2 Department of Odontostomatological and Maxillofacial Sciences, Sapienza—University of Rome, 00161 Rome, Italy
* Correspondence: giacomo.visioli@uniroma1.it (G.V.); alessandro.lambiase@uniroma1.it (A.L.)

Abstract: Italy is a natural corridor for entry into Europe, receiving thousands of refugees and migrants needing socio-economic and health assistance yearly. Impaired vision due to eye disease is estimated to affect at least 2.2 billion people worldwide, especially in this underprivileged population. To overcome this deep disparity, new intervention strategies, such as the PROTECT project, were planned with the aim of assessing, in the context of the head–neck area, the eye health in vulnerable applicants and holders of international protection. A total of 3023 migrants were involved in the project. Demographic factors and eye history were collected using a questionnaire. Using portable diagnostic instruments, an eye screening including monocular visual acuity, intraocular pressure, anterior segment, and ocular fundus was performed. The mean age was 31.6 ± 13.1 years and more than 50% underwent the first eye evaluation. Vision impairment was claimed by 16.6% of subjects and the most frequent diseases diagnosed were: refractive errors (11%), strabismus (6%), red eye (6%), cataract (5.3%), and ocular hypertension (1%). Retinal alterations were observed in 5% of migrants. The PROTECT project allows us to increase the accessibility of head–neck disease prevention care. Moreover, our results confirm the utility of an eye screening assessment for early identification of the most relevant and preventable ocular diseases, especially in disadvantaged populations.

Keywords: impaired ocular vision; ocular diseases; screening; migrants; vulnerability; cooperation; hospitality

Citation: Bruscolini, A.; Visioli, G.; Marenco, M.; Cherubini, V.; Comberiati, A.M.; Palaia, G.; Ralli, M.; Ottolenghi, L.; Lambiase, A.; Polimeni, A. Eye Health Screening in Migrant Population: Primary Care Experience in Lazio (Italy) from the PROTECT Project. *Appl. Sci.* **2023**, *13*, 3618. https://doi.org/10.3390/app13063618

Academic Editor: Alexandros A. Lavdas

Received: 7 February 2023
Revised: 1 March 2023
Accepted: 9 March 2023
Published: 12 March 2023

1. Introduction

Eye health and vision have profound and widespread implications for many aspects of life, health, sustainable development, and the economy. Currently, many people, families, and populations continue to suffer the consequences of vision impairment and blindness [1,2].

Eye conditions causing low vision and blindness are remarkably common. It is estimated that almost all individuals will experience impaired vision or an eye condition during the life course and require eye care services. According to the latest World Health Organization (WHO) Report on Vision, at least 2.2 billion people worldwide have impaired vision due to either eye diseases (age-related macular disease, cataract, diabetic retinopathy, glaucoma) or uncorrected refractive errors. In at least 1.1 billion of these cases, vision impairment could have been prevented or has yet to be addressed with associated highly cost-effective interventions [1]. By 2050, this figure is expected to rise to 1.8 billion [1,3]. Accordingly, the leading causes of vision impairment are uncorrected refractive errors and cataracts [3,4]. Since these conditions can be effectively corrected or treated, early detection may improve clinical outcomes and reduce disease burden on patients and society. In addition, recent data showed an increasing prevalence of pediatric uncorrected refractive error worldwide which may cause amblyopia. It is the most common cause of impaired monocular visual acuity in children, and is treatable if diagnosed early with great functional success [5].

The burden tends to be greater in low- and middle-income settings among older people, in women, and in rural and disadvantaged communities due to fewer opportunities to access the most essential eye care services [6,7]. Due to its central position in the Mediterranean Sea, Italy represents an interesting country to explore this issue as it constantly receives migrants coming from different countries [8].

At the end of 2021, the total number of accommodated asylum seekers and beneficiaries of international protection in Italian centers such as CAS (Emergency Accommodation Centre) and SAI (System of Accommodation and Integration) was 78,001. The region of Lazio accommodated the fourth largest number of migrants in Italy after Lombardy, Emilia Romagna and Piedmont (12.6%, 10% and 9%, respectively), with a total of 6813 (8.6%) [9]. Expectably, in the regions with the highest amount of asylum seekers, socio-economic integration and health care was a tough challenge [10,11].

In Italy, according to Italian legislation, within the framework of the National Plan for the Integration of Internationally Protected Persons, the planning of integration policies, including health care, is a regional prerogative [12]. Local authorities and not-for-profit organizations may request funding from national or EU funds for projects designed to improve the integration of the migrant population including health screening. In this regard, the "Migration and Integration Asylum Fund 2014–2020" (AMIF) is a financial instrument established by EU Regulation No. 516/2014 with the objective of promoting an integrated management of migration flows supporting all aspects of the phenomenon: asylum, integration and return [13]. Through the AMIF fund, the Italian Ministry of Internal Affairs financed the PROTECT project. PROTECT ("Patologie del distRettO TEsta-Collo nei minori migranTi. Dalla formazione degli operatori alla diagnosi precoce e presa in carico del paziente: Network Odontoiatrico, Oftalmologico, Otorinolaringoiatrico e Maxillo-Facciale") is a multidisciplinary project of secondary prevention of ophthalmic, oral and dental disease in migrant and vulnerable populations. It was carried out by the Hospital-University Polyclinic Umberto I (Department Integrated Head–neck), in collaboration with the Departments of Sense Organs and the Department of Odontostomatological and Maxillofacial Sciences of the Sapienza—University of Rome, Lazio [14]. PROTECT project results from oral screening have been already made available and dental pathologies turned out to be widespread and often neglected in asylum seekers, with a high relative abundance of malocclusions and carious lesions. Furthermore, results from this screening have confirmed the crucial role of prevention in a low-income setting to reduce hospitalizations and worsening problems [15].

Regarding eye screening, the role of prevention of ocular pathologies in the general population had been extensively discussed, with a relevant reduction of complications especially in diabetic patients [16,17]. Many studies have previously investigated visual health and visual healthcare access in refugees or displaced persons, with a prevalence of blindness ranging from 1.6 to 26.2% and a high prevalence of ocular infectious diseases such as trachoma [18]. However, scarce real-world data regarding ocular health in the setting of asylum seekers or beneficiaries of international protection are currently available.

The aim of this paper is to describe the results of the ophthalmological screening from the PROTECT project and to discuss the social and medical need of early diagnosis and vision impairment management, especially in vulnerable populations.

2. Materials and Methods

This is a national cross-sectional study ("PROTECT" project) conducted among the refugee and migrant population present in the Lazio region, Italy, from February 2018 to September 2021. The vision health of 3023 participants was assessed in a network of 53 cultural associations and reception centers. This study adheres to the tenets of the Declaration of Helsinki for research involving human subjects, and the "PROTECT" project was approved by the Ethics Committee of the Department of Odontostomatological and Maxillofacial Sciences of the Sapienza—University of Rome (Protocol identifying number: 0000839 on 2 October 2018). The ocular screening was performed by trained ophthalmologists and optometrists of the Department of Sense Organs of the Sapienza University —Rome at the clinics of the Policlinico Umberto I University Hospital or at the reception centers using mobile units.

2.1. Data Collection

A screening folder was prepared by the project staff to identify subjects with eye diseases requiring additional specialized care. Red or yellow codes were used to label the severity of the disease. The screening data were collected via a health survey in order to achieve details on basic demographics, time since the last eye examination, and state of eye health. Using a binocular portable refractometer device (EsaVision Adaptica 2 WIN), all participants underwent ophthalmological examination for assessment of the following clinical data: objective refraction, ocular motility, and presence of the major amblyogenic factors such as anisometropia, anisocoria, and media opacities. Visual acuity was measured using the subjective refraction method with a Snellen chart, and patients with visual acuity of less than 20/40 were referred for further visual examination in a specialist clinic. Intraocular pressure was evaluated with a rebound tonometer (iCare TA01i tonometer) [19]. Moreover, the presence/absence of ocular surface inflammation (red eye) and macular area disease were recorded in the medical sheet using Smartscope PRO (Optomed, Finland) handheld, a lightweight and affordable fundus camera [20].

2.2. Questionnaire

The questionnaire was applied individually through an interview with the help of a linguistic mediator. It included items related to sociodemographic variables and ocular health as shown in Table 1.

Table 1. Questionnaire applied at the time of enrollment to assess ocular health.

1. Sociodemographic Variable	**Answers (Yes/NO/NA)**
Full name	
Gender	
Age (years)	
Country of origin	
Pre-disposing conditions/family history	
2. Ocular Health-Related Behaviors	
Do you have ever had an eye evaluation?	
Do you have previous eye injuries?	
Do you use eyeglasses?	
Do you have visual impairment?	
Do you have ocular pain right now?	

2.3. Clinical Examinations

All subjects underwent the following evaluations: (i) autorefractor exam; (ii) eye motility and stereopsis; (iii) distance and near visual acuity using the Snellen Tumbling E Charts at distance and near; (iv) intraocular pressure with no contact tonometry; and (v) segment anterior and macular photograph.

In order to detect the possible cause of preventable vision impairment, the variables analyzed were: type of refractive error, number (%) of the subject with visual acuity <20/40 at distance, and cataract. In addition, we assessed the presence of ocular signs of disease requiring urgent care (red card), such as infective or neoplastic disease, or minor urgence (yellow card), such as glaucoma suspicion (intraocular pressure > 20 mmHg) or retinal abnormalities. Patients who needed further ophthalmologic evaluation after the screening received assistance by joining the healthcare network of the Policlinico Umberto I Hospital.

2.4. Data Analysis

Data were analyzed using standard statistical analysis software (version 20.0, Statistical Package for the Social Sciences, IBM Corporation, Armonk, NY, USA). A database was developed using Excel (Microsoft, Redmond, WA, USA). Descriptive statistics including mean ± SD values and percentage were calculated for each variable.

3. Results

A total of 3023 patients were involved in the project PROTECT. The mean age was 31.6 ± 13.1 years, and among all the subjects 2058 were male (68.1%) and 965 were women (31.9%). The geographical origin of the subjects is illustrated in Figure 1.

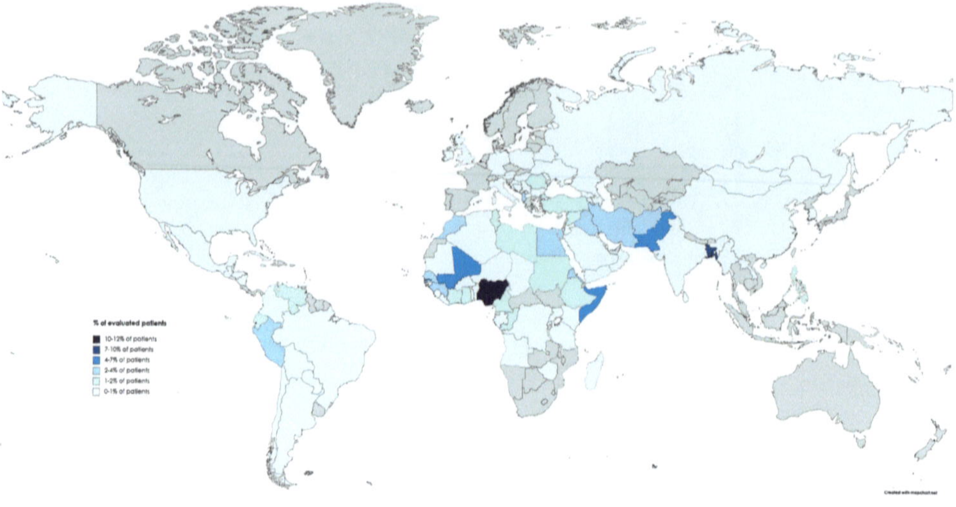

Figure 1. World map representing the geographical origin of the screened subjects. Of note, 10.3% of the screened migrants were Nigerians, 7.6% Bangladeshi and 5.6% Pakistani.

The questionnaire replies showed that 50.4% of subjects had never had an eye examination, with an additional number who were unable to provide a definitive response (8.9%). Most participants (86.2%) were unaware of their family's medical history. In addition, most of the patients claimed vision impairment (16.6%), ocular pain (6.6%), and previous history of ocular injury (5.2%). Finally, the use of eyeglasses was reported by 12.0% of the sample. A summary chart is reported in Figure 2.

The most frequent eye alteration in our population was a refractive error; in fact, this was found in 11% of all participants, and it was distributed between myopia (9.8%) and hyperopia (1.2%) among all subjects.

Visual acuity evaluation showed that 9.6% of subjects had uncorrected vision acuity (UCVA) less than 20/40 associated with ocular motility alterations and stereopsis loss.

The overall prevalence of other eye diseases including strabismus (6%), red eye (6%), cataract (5.3%), and IOP values above 20 mmHg (1%) is represented in Figure 3.

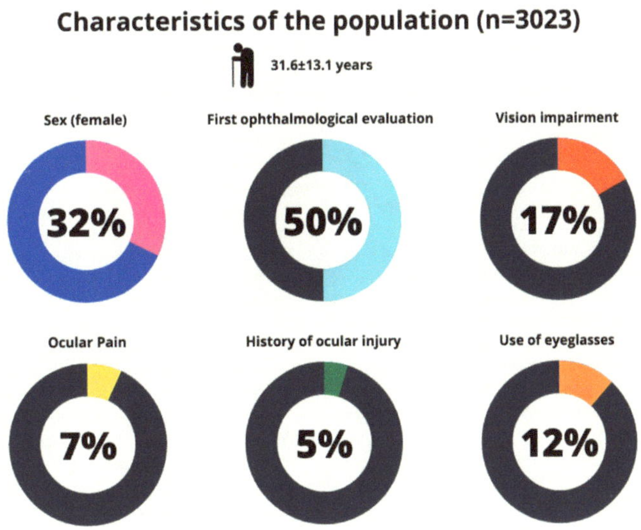

Figure 2. Main characteristics of the screened population. Of note, more than 50% of migrants underwent the first ophthalmological evaluation.

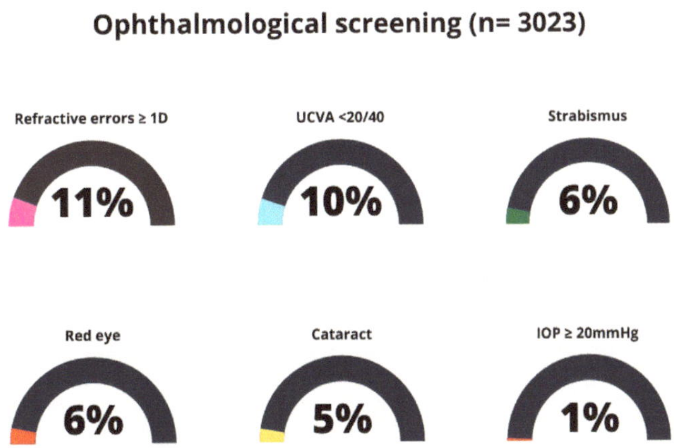

Figure 3. Principal findings of the ophthalmological screening within the PROTECT project. Of note, refractive error represented the main cause of vision impairment.

The frequency of ocular fundus changes was 5%, with several possible causes shown in Figure 4. In particular, ten subjects had isolated cotton wool spots and four participants showed some microaneurysms without a documented history of diabetes that required further examination in a specialist hospital. Indeed, 60% of patients with optic nerve abnormalities had bilateral symmetric changes in terms of cup disc ratio.

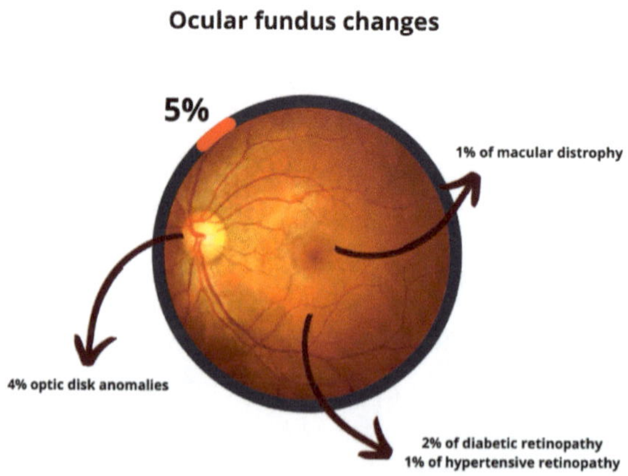

Figure 4. Principal retinal findings of the screened migrants within the PROTECT project. Of note, in certain cases, multiple findings coexisted in the same patient.

In addition, during ophthalmological screening 18 of 51 subjects with red eyes showed signs of acute conjunctivitis and were sent for urgent care to a specialist eye clinic.

More than 30% of the total population screened were referred to the Policlinico Umberto I University Hospital Medical Center to carry out further investigations such as: correction of refraction error with eyeglasses, strabismus treatment, management of ocular surface inflammation, cataract surgery evaluation and investigations to identify glaucoma early.

4. Discussion

Screening for vision impairment is a highly effective way to prevent long-term disability, as recently reiterated by the WHO [2]. A few studies have addressed eye health among migrant populations, especially in Italy, and the PROTECT project is the first Italian project tailored to the head and neck health of a wide sample of fragile populations that live in reception centers and have several different demographic origins [14].

Indeed, very few screening tools have been validated for diagnosis accuracy among migrants overall, and several heterogeneous data have been reported on the ocular health of migrants in the word, highlighting the lack of proper medical care for this fragile population. In fact, while a reduction in visual impairment in high-income countries has been described, migrants and vulnerable populations that live in these European countries experience obstacles to health care access. As shown in our study, about 50% of migrants carried out their first ophthalmological visit during this screening project and about 9% of subjects could not report if they had a previous eye exam. These data, in line with the previous evidence, suggest a potential increased risk of amblyopia [21]. This defect of vision becomes irreversible in adulthood, affecting the quality of life, and interfering with the ability to socialize and work. Recent studies showed that screening for amblyopia is cost-effective, but highly related to long-term impacts of unilateral loss that are common in migrant populations.

In fact, analyzing the answers to the questionnaire, the presence of vision impairment or ocular pain (the main reasons for a request for an eye examination in the Italian population) was reported by 17% and 7% of subjects, respectively.

The PROTECT project result showed that an easy and fast eye screening with portable technologies is able to highlight the main causes of visual problems and to plan specific management of the patient with different levels of urgency [22]. In addition, the strategy of our screening, based on a non-invasive assessment and in many cases carried out directly at the center of reception, has allowed the obtaining of good compliance by these vulnerable subjects often reticent to visits and treatment [23].

According to evidence, the eye screening results indicate that uncorrected refractive error was the more frequent finding associated with visual impairment, and patients were referred within 30 days (yellow card) at the optometrist clinic to evaluate the prescription of eyeglasses and associated ocular motility alterations. A special agreement with a large private company for the provision of spectacles for refugees is fundamentally advisable to complete the management of vision impairment of this disadvantaged population.

Despite the sample's young average age, cataracts were the second most frequent cause of vision loss, likely related to traumatic or dysmetabolic diseases, and a second appointment in the specialist eye hospital was booked to follow up and manage the patients [24]. Interestingly, by using simple and rapid eye screening programs in the reception centers, the diagnosis of acute pathologies such as ocular hypertension and conjunctivitis, was performed, and urgent treatment or further investigations in specialist vision care were started (red card).

The few ocular fundus changes requiring further investigation were managed by the retinal specialist (yellow card) to diagnose dysmetabolic pathologies early, such as diabetes or hypertension, or to follow up on choroidal nevus. Optic disc abnormalities suggesting potential glaucoma disease showed a prevalence of 4%, although it is important to remember that the diagnosis of this pathology needs an evaluation of the optic disc by ophthalmoscopy exam with pharmacological mydriasis and instrumental examinations, such as the visual field carried and optical coherence tomography (OCT) [25]. Although the role of the digital fundus camera in glaucoma diagnosis during screening programs remains controversial due to the quality of photographs or algorithm performance, in our opinion, for the purpose of the PROTECT project, this diagnostic procedure proved to be a crude but useful evaluation to highlight the possible subjects at risk and to guide clinicians, as has been suggested recently [26–29].

In contrast to other studies, no age-related macular diseases were observed, and we suppose that these data are related to the average young age of the sample [30]. Our screening could not assess the presence of any retinal peripheral lesions at risk of retinal detachment. The weaknesses of this study are related to average age and heterogeneity of the sample. The young age is a great bias that influences the results of our study; in fact, we diagnosed a very low number of AMD and cataracts, which are usually related to age. In addition, scarce and limited evidence on ocular disorder prevalence in young immigrants with different geographical provenience has been published, as noted in a paper by Global Burden of Disease Study Blindness and Vision Impairment Collaborators [31], making it more difficult to assess the value of our results in immigration-related conditions.

An important enrichment in terms of care for these disadvantaged populations could also come from the use of teleophthalmology. In fact, also during this project, the restrictions linked to the COVID-19 pandemic led to the development of remote consultation systems to maintain a mode of assistance with the reception centers in the most critical periods [32].

5. Conclusions

The multidisciplinary approach to screening head and neck diseases has allowed the development of a network of care essential to meet the different medical needs of a fragile and neglected population with the aim to improve quality of life and socio/economic integration. A permanent head–neck care service for all fragile populations reaching our country is desirable and represents an innovative prevention service for the national health system.

In the future, the validation of clinical smartphone applications in ophthalmology will help to further improve the operation and cost of teleophthalmology in the field of prevention, so as to carry out not only remote screening programs but also monitor the course of diseases diagnosed with periodic follow-up. Indeed, it would be desirable to have a permanent channel of the first consultation with institutions hosting fragile people in order to create a rapid, effective and affordable system for the national health system.

Author Contributions: Conceptualization, A.B. and A.P.; methodology, A.B and A.L.; software, G.P.; validation, A.B., M.M. and A.L.; formal analysis, G.V.; investigation, A.B., M.M., A.M.C. and V.C.; resources, A.P. and L.O.; data curation, M.R.; writing—original draft preparation, A.B.; writing—review and editing, G.V.; visualization, G.V.; supervision, A.L.; project administration, A.P. and L.O.; funding acquisition, A.P. All authors have read and agreed to the published version of the manuscript.

Funding: This research received external funding from the Asylum, Migration and Integration Fund (AMIF) of the European Union, according to the code PROG-2010.

Institutional Review Board Statement: The study was conducted in accordance with the Declaration of Helsinki, and approved by the Ethics Committee of Sapienza—University of Rome (Protocol identifying number: 0000839 and approved on 2 October 2018).

Informed Consent Statement: Informed consent was obtained from all subjects involved in the study.

Data Availability Statement: The data presented in this study are available on reasonable request from the corresponding authors, after board approval. The data are not publicly available due to informed consent restrictions.

Conflicts of Interest: The authors declare no conflict of interest.

References

1. World Health Organization. World Report on Vision. Available online: https://www.iapb.org/wp-content/uploads/2020/09/world-vision-report-accessible1.pdf (accessed on 15 January 2023).
2. Abdulhussein, D.; Abdul Hussein, M. WHO Vision 2020: Have We Done It? *Ophthalmic Epidemiol.* **2022**, 1–9. [CrossRef]
3. Burton, M.J.; Ramke, J.; Marques, A.P.; Bourne, R.R.A.; Congdon, N.; Jones, I.; Ah Tong, B.A.M.; Arunga, S.; Bachani, D.; Bascaran, C.; et al. The Lancet Global Health Commission on Global Eye Health: Vision beyond 2020. *Lancet Glob. Health* **2021**, *9*, e489–e551. [CrossRef] [PubMed]
4. Ackland, P.; Resnikoff, S.; Bourne, R. World Blindness and Visual Impairment: Despite Many Successes, the Problem Is Growing. *Community Eye Health J.* **2018**, *30*, 71–73.
5. Hu, B.; Liu, Z.; Zhao, J.; Zeng, L.; Hao, G.; Shui, D.; Mao, K. The Global Prevalence of Amblyopia in Children: A Systematic Review and Meta-Analysis. *Front. Pediatr.* **2022**, *10*, 819998. [CrossRef]
6. Keel, S.; Lingham, G.; Misra, N.; Block, S.; Bourne, R.; Calonge, M.; Cheng, C.-Y.; Friedman, D.S.; Furtado, J.M.; Khanna, R.; et al. Toward Universal Eye Health Coverage—Key Outcomes of the World Health Organization Package of Eye Care Interventions: A Systematic Review. *JAMA Ophthalmol.* **2022**, *140*, 1229–1238. [CrossRef]
7. Palaia, G.; Bruscolini, A.; Ralli, M. The Prevention and Treatment of Medical Diseases in Vulnerable Populations. *Appl. Sci.* **2022**, *12*, 10406. [CrossRef]
8. Campomori, F.; Ambrosini, M. Multilevel Governance in Trouble: The Implementation of Asylum Seekers' Reception in Italy as a Battleground. *Comp. Migr. Stud.* **2020**, *8*, 22. [CrossRef]
9. AIDA. Asylum Information Database. Country Report: Italy. Available online: www.asylumineurope.org (accessed on 15 January 2023).
10. Napoli, C.; Riccardo, F.; Declich, S.; Dente, M.G.; Pompa, M.G.; Rizzo, C.; Rota, M.C.; Bella, A. An Early Warning System Based on Syndromic Surveillance to Detect Potential Health Emergencies among Migrants: Results of a Two-Year Experience in Italy. *Int. J. Environ. Res. Public Health* **2014**, *11*, 8529–8541. [CrossRef]
11. Lauritano, D.; Moreo, G.; Carinci, F.; Campanella, V.; della Vella, F.; Petruzzi, M. Oral Health Status among Migrants from Middle- and Low-Income Countries to Europe: A Systematic Review. *Int. J. Environ. Res. Public Health* **2021**, *18*, 12203. [CrossRef]
12. World Health Organization. Italy: Health System Review. Available online: https://apps.who.int/iris/handle/10665/141626 (accessed on 23 January 2023).
13. Migration and Home Affairs. Asylum, Migration and Integration Fund (2014–2020). Available online: https://home-affairs.ec.europa.eu/funding/asylum-migration-and-integration-funds/asylum-migration-and-integration-fund-2014-2020_en (accessed on 15 January 2023).
14. The PROTECT Project. Available online: https://www.famiprotect.it/ (accessed on 15 January 2023).
15. Palaia, G.; Bossù, M.; Pranno, N.; Ralli, M.; Bruscolini, A.; Capocci, M.; Gaimari, G.; Macali, F.; Corridore, D.; Cotticelli, L.; et al. Oral Pathologies in Migrants: The Experience of the PROTECT Project in 3023 Patients. *Appl. Sci.* **2022**, *12*, 12621. [CrossRef]
16. Scanlon, P.H. The Contribution of the English NHS Diabetic Eye Screening Programme to Reductions in Diabetes-Related Blindness, Comparisons within Europe, and Future Challenges. *Acta Diabetol.* **2021**, *58*, 521–530. [CrossRef] [PubMed]
17. White, N.H.; Pan, Q.; Knowler, W.C.; Schroeder, E.B.; Dabelea, D.; Chew, E.Y.; Blodi, B.; Goldberg, R.B.; Pi-Sunyer, X.; Darwin, C.; et al. The Effect of Interventions to Prevent Type 2 Diabetes on the Development of Diabetic Retinopathy: The DPP/DPPOS Experience. *Diabetes Care* **2022**, *45*, 1640–1646. [CrossRef]
18. Bal, S.; Duckles, A.; Buttenheim, A. Visual Health and Visual Healthcare Access in Refugees and Displaced Persons: A Systematic Review. *J. Immigr. Minor. Health* **2019**, *21*, 161–174. [CrossRef] [PubMed]

19. Nakakura, S.; Mori, E.; Fujio, Y.; Fujisawa, Y.; Matsuya, K.; Kobayashi, Y.; Tabuchi, H.; Asaoka, R.; Kiuchi, Y. Comparison of the Intraocular Pressure Measured Using the New Rebound Tonometer Icare Ic100 and Icare TA01i or Goldmann Applanation Tonometer. *J. Glaucoma* **2019**, *28*, 172–177. [CrossRef] [PubMed]
20. Gosheva, M.; Klameth, C.; Norrenberg, L.; Clin, L.; Dietter, J.; Haq, W.; Ivanov, I.V.; Ziemssen, F.; Leitritz, M.A. Quality and Learning Curve of Handheld versus Stand-Alone Non-Mydriatic Cameras. *Clin. Ophthalmol.* **2017**, *11*, 1601–1606. [CrossRef]
21. Fricke, T.R.; Holden, B.A.; Wilson, D.A.; Schlenther, G.; Naidoo, K.S.; Resnikoff, S.; Frick, K.D. Coût Global de Correction d'une Déficience Visuelle Induite Par Une Erreur de Réfraction Non Corrigée. *Bull. World Health Organ.* **2012**, *90*, 728–738. [CrossRef]
22. Labkovich, M.; Paul, M.; Kim, E.; Serafini, R.A.; Lakhtakia, S.; Valliani, A.A.; Warburton, A.J.; Patel, A.; Zhou, D.; Sklar, B.; et al. Portable Hardware & Software Technologies for Addressing Ophthalmic Health Disparities: A Systematic Review. *Digit. Health* **2022**, *8*, 20552076221090042. [CrossRef]
23. Németh, J.; Szabó, D.; Tóth, G.; Sándor, G.; Lukács, R.; Pék, A.; Szalai, I.; Papp, A.; Resnikoff, S.; Limburg, H. Feasibility of the Rapid Assessment of Avoidable Blindness with Diabetic Retinopathy Module (RAAB+DR) in Industrialised Countries: Challenges and Lessons Learned in Hungary. *Ophthalmic Epidemiol.* **2018**, *25*, 273–279. [CrossRef]
24. Nwosu, S.N.N. Ocular Problems of Young Adults in Rural Nigeria. *Int. Ophthalmol.* **1998**, *22*, 259–263. [CrossRef]
25. Wagner, I.V.; Stewart, M.W.; Dorairaj, S.K. Updates on the Diagnosis and Management of Glaucoma. *Mayo Clin. Proc. Innov. Qual. Outcomes* **2022**, *6*, 618–635. [CrossRef]
26. Upadhyaya, S.; Agarwal, A.; Rengaraj, V.; Srinivasan, K.; Newman Casey, P.A.; Schehlein, E. Validation of a Portable, Non-Mydriatic Fundus Camera Compared to Gold Standard Dilated Fundus Examination Using Slit Lamp Biomicroscopy for Assessing the Optic Disc for Glaucoma. *Eye* **2021**, *36*, 441–447. [CrossRef]
27. Yogesan, K.; Constable, I.J.; Barry, C.J.; Eikelboom, R.H.; McAllister, I.L.; Tay-Kearney, M.L. Telemedicine Screening of Diabetic Retinopathy Using a Hand-Held Fundus Camera. *Telemed. J.* **2000**, *6*, 219–223. [CrossRef] [PubMed]
28. Ertel, M.K.; Kahook, M.Y.; Capitena Young, C.E. The Future Is Now: Incorporating Telemedicine into Glaucoma Care. *Curr. Ophthalmol. Rep.* **2021**, *9*, 88–95. [CrossRef]
29. Myers, J.S.; Fudemberg, S.J.; Lee, D. Evolution of Optic Nerve Photography for Glaucoma Screening: A Review. *Clin. Exp. Ophthalmol.* **2018**, *46*, 169–176. [CrossRef]
30. Owsley, C.; McGwin, G.; Lee, D.J.; Lam, B.L.; Friedman, D.S.; Gower, E.W.; Haller, J.A.; Hark, L.A.; Saaddine, J. Diabetes Eye Screening in Urban Settings Serving Minority Populations: Detection of Diabetic Retinopathy and Other Ocular Findings Using Telemedicine. *JAMA Ophthalmol.* **2015**, *133*, 174–181. [CrossRef]
31. GBD 2019 Blindness and Vision Impairment Collaborators; Vision Loss Expert Group of the Global Burden of Disease Study. Causes of blindness and vision impairment in 2020 and trends over 30 years, and prevalence of avoidable blindness in relation to VISION 2020: The Right to Sight: An analysis for the Global Burden of Disease Study. *Lancet Glob. Health* **2021**, *9*, e144–e160. [CrossRef]
32. Chong, J.C.; Tan, C.H.N.; Chen, D.Z. Teleophthalmology and Its Evolving Role in a COVID-19 Pandemic: A Scoping Review. *Ann. Acad Med. Singap.* **2021**, *50*, 61–76. [CrossRef] [PubMed]

applied sciences

Review

Childhood Uveitic Glaucoma: Complex Management in a Fragile Population

Valeria Iannucci [1,†], Priscilla Manni [1,*,†], Giulia Mecarelli [1], Sara Giammaria [2], Francesca Giovannetti [1], Alessandro Lambiase [1,*] and Alice Bruscolini [1]

1 Department of Sense Organs, Sapienza University of Rome, 00185 Rome, Italy
2 IRCCS-Fondazione Bietti, 00198 Rome, Italy
* Correspondence: priscilla.manni@uniroma1.it (P.M.); alessandro.lambiase@uniroma1.it (A.L.)
† These authors contributed equally to this work.

Abstract: Uveitic glaucoma (UG) is a potentially blinding complication of intraocular inflammation and is one of the most common causes of secondary glaucoma in pediatric ophthalmology. Overall management of UG is often challenging and requires a multidisciplinary assessment and careful follow-up. The overlap with steroid-induced glaucoma (SIG) is quite common, as well as the failure of medical and surgical therapy; nevertheless, few recent papers have dealt with this topic. We review the features and the clinical approach to UG in childhood, discussing the treatments available in the pediatric population.

Keywords: uveitis and intraocular pressure; pediatric uveitis; uveitic glaucoma; pediatric uveitic glaucoma; steroid and glaucoma; pediatric glaucoma drugs; glaucoma surgery

check for updates

Citation: Iannucci, V.; Manni, P.; Mecarelli, G.; Giammaria, S.; Giovannetti, F.; Lambiase, A.; Bruscolini, A. Childhood Uveitic Glaucoma: Complex Management in a Fragile Population. *Appl. Sci.* **2023**, *13*, 2205. https://doi.org/10.3390/app13042205

Academic Editor: Gaetano Isola

Received: 19 January 2023
Revised: 6 February 2023
Accepted: 7 February 2023
Published: 9 February 2023

1. Introduction

Childhood glaucoma is a heterogeneous group of disorders often associated with severe visual loss affecting individuals under the age of 16 years old (UK, Europe, UNICEF) or under 18 years old (USA). It affects more than 300,000 children worldwide and accounts for 5% of the causes of blindness in the pediatric population [1,2].

The Childhood Glaucoma Research Network (CGRN) proposed a novel classification system to unify nomenclature in childhood glaucoma [3]. According to this classification, the diagnosis of childhood glaucoma requires the presence of two or more of the following criteria:

- Intraocular pressure (IOP): >21 mmHg;
- Visual field defects consistent with glaucomatous optic neuropathy;
- Progressive myopia or myopic shift with increased axial length (AL);
- Presence of Haab striae or increased corneal diameter;
- Progressive increase in cup-disc ratio (C/D), cup-disc asymmetry of 0.2 or more between both eyes and/or focal rim thinning.

The CGRN classification system identifies primary or secondary forms of childhood glaucoma. Primary glaucoma can be further subdivided into Primary Congenital Glaucoma (PCG) and Juvenile Open-Angle Glaucoma (JOAG).

Secondary glaucoma includes four categories: glaucoma following cataract surgery, glaucoma associated with non-acquired systemic disease or syndrome, glaucoma associated with non-acquired ocular anomalies and glaucoma related to acquired conditions such as ocular trauma or uveitis.

Glaucoma secondary to uveitis (UG) is one of the most common causes of secondary glaucoma in the pediatric population. Despite this, few recent papers have dealt with this topic.

In childhood, anterior uveitis is the prevalent cause of secondary glaucoma [4,5]. Moreover, children with uveitis have a higher risk of developing complications, including glaucoma, compared to adults with uveitis [6].

Whilst pediatric uveitis represents 5–10% of all uveitis cases, visual loss is more prevalent and severe, resulting in no light perception in one-third of patients, according to Kanski et al. [7]. Visual outcomes depend on the anatomical location and duration of the uveitis, age of presentation, management of intraocular inflammation and the prescribed corticosteroid therapy [8].

The British Infantile and Childhood Glaucoma (BIG) eye study shows that 5.3–19% of all childhood glaucoma cases in the United Kingdom are caused by uveitis [9].

Paroli et al., reported in a previous work a 25% prevalence of uveitic glaucoma in children affected by uveitis [10]. In a retrospective analysis of 182 pediatric uveitis patients, secondary glaucoma was reported in 48 patients (26.23%) with female predominance (F:M, 29:19) [4].

Complications involving the eye structures in children uveitis differ from the ones in adult patients uveitis in terms of etiology, prognosis and prevalence [6]. Ocular complications occur in around 76% of children with uveitis. Pediatric uveitis is often misdiagnosed, and this increases the onset of well-known complications, such as cataracts, band keratopathy, glaucoma, and cystoid macular edema but also of specific complications for the pediatric population, such as amblyopia [8]. De Boer et al., reported glaucoma as the second most common complication (19%) after cataracts and one of the most common causes of severe visual impairment in children with uveitis [11]. The therapeutic approach of UG is still challenging due to poor randomized controlled trials (RCTs) in this pediatric population and the lack of universally agreed guidelines.

The aim of this paper is to describe the clinical features of uveitic glaucoma in a fragile population, such as pediatric patients, and to provide an update on etiopathogenesis and the current therapeutic approach.

2. Classification and Clinical Features of UG in Childhood

The Standardization of Uveitis Nomenclature (SUN) system [12] classifies uveitis according to the anatomic location of inflammation into anterior (iris and the anterior ciliary body), intermediate (posterior ciliary body and vitreous), posterior (retina and/or choroid), and panuveitis (all structures are affected). In addition, this classification provides information about the disease, such as onset (sudden or insidious), duration [limited ($\leq$3 months) or persistent (>3 months)], and course (acute, recurrent, or chronic).

Here we report the features of pediatric diseases and infections which are more frequently related to uveitic glaucoma, according to the SUN system. Among the anterior uveitis, we reported the Juvenile Idiopathic Arthritis JIA-associated uveitis, the herpetic anterior uveitis and tubulointerstitial Nephritis and Uveitis (TINU).

Juvenile Idiopathic Arthritis (JIA) is now defined as arthritis starting before the age of 16 years and lasting for 6 weeks or more. The disease activity reduces around 9 years of age and peaks around puberty. [13,14]. JIA-associated uveitis is a chronic non-granulomatous anterior uveitis, and in 3–7% of cases, it can precede the diagnosis of arthritis [15]. Unlike other forms of anterior uveitis, JIA-associated uveitis has an insidious onset, either unilateral or, more often, bilateral, and it is usually asymptomatic until sight-threatening complications arise. About three-quarters develop chronic inflammation with a high incidence of complications, such as cataracts, glaucoma, band keratopathy, and persistent cystoid macular edema [16]. The severity and chronicity of JIA-associated uveitis vary markedly. Young age at onset of uveitis, male genders, presence of synechiae at presentation and uveitis occurring before arthritis appear to be the most significant risk factors for a severe course of uveitis and developing complications [5]. A minority have mild, self-limiting uveitis, requiring only short-term topical steroids [17].

Herpetic anterior uveitis is another form potentially responsible for secondary glaucoma in children. Infectious etiologies constitute 13% of all pediatric uveitis [18]. Viruses implicated are Herpes Simplex Virus (HSV) type 1 and type 2, Varicella Zoster Virus (VZV), and Cytomegalovirus (CMV). They can cause acute unilateral non-granulomatous anterior uveitis, often associated with increased intraocular pressure, but they can also cause granulomatous anterior uveitis in chronic stages. Although rare in children, when the infection involves the retina, Acute Retinal Necrosis (ARN) may occur, causing a devastating reduction in visual acuity.

Tubulointerstitial Nephritis and Uveitis (TINU) is a multisystemic autoimmune disease involving uvea and renal tubules and may be triggered by various factors, such as drugs or infections. Its incidence is highest at 15 years of age, but all age groups can be affected. In TINU, the uveitis is anterior and usually involves both eyes r; however, posterior or intermediate uveitis can occur as well.

Among the intermediate uveitis in childhood, pars planitis (PP), early-onset sarcoidosis (Blau syndrome), and juvenile multiple sclerosis (MS) are worth mentioning [19–21]. PP is an idiopathic type of intermediate uveitis accounting for 5–26.7% of all pediatric uveitis and is a diagnosis of exclusion [21]. Blau Syndrome is a systemic inflammatory granulomatous disease that occurs in children under 5 years. It is characterized by uveitis, arthritis (mostly involving the knee and the wrist), and skin lesions, such as erythema nodosum, skin rash, vasculitis etc. Juvenile multiple sclerosis (MS) is rare in children, with an incidence in the literature ranging from 2.7% to 10.5% of the general MS group [22].

Posterior uveitis includes inflammation of the choroid with or without retinal involvement. In children, as in adults, the most common cause is toxoplasmosis [23,24]. In 70–80% of cases, it appears as unilateral focal necrotizing retinochoroiditis with focal vitritis. The anterior segment might be secondarily involved, presenting as granulomatous anterior uveitis and high intraocular pressure. Ocular toxoplasmosis more commonly results from the reactivation of congenital disease, where new active satellite lesions appear next to an atrophic scar with hyperpigmented borders. The absence of scarring lesions suggests acquired disease. Congenital toxoplasmosis occurs due to the passage of T.gondii by the transplacental route. The transmission risk is highest during late pregnancy, but most severe forms occur during the first trimester with the classic triad of retinochoroiditis, cerebral calcifications and seizures [25,26].

Other forms of posterior uveitis are infrequent in childhood. Vogt–Koyanagi–Harada (VKH) is a multisystem inflammatory autoimmune rare disease affecting the eyes, ears, brain, skin and hair [27,28]. It is more frequent in ethnicities with higher pigmentation, such as Asians, Middle Easterners, Hispanics, and Native Americans. The incidence of VKH in the pediatric population is ethnicity-dependent, varying between 0.5 and 3% of all pediatric uveitis [29]. The diagnosis is usually delayed in children compared to adults, so associated ocular complications, including glaucoma, occur more frequently. Therefore, visual outcomes in VKH are usually worse in the pediatric population [30].

Another entity associated with posterior uveitis in children is Behçet's disease (BD). BD is a multisystemic vasculitis that is less common in pediatric patients than adults, and it is characterized by recurrent mucocutaneous ulcers affecting the oral cavity and the genital area. Pediatric BD onset is around 10–15 years of age, and ocular involvement usually occurs within 2–3 months. It includes bilateral recurrent panuveitis with retinal vasculitis and persisting vitritis. Iridocyclitis, episcleritis, retinitis, retinal hemorrhages, optic nerve edema, and cystoid macular edema can also be found [31].

Data on the relative risk of developing glaucoma in pediatric patients according to the type of uveitis [4] are limited and extremely heterogeneous (Table 1).

Table 1. Type of uveitis and prevalence of pediatric uveitis with the corresponding relative risk (RR) of UG. JIA: Juvenile Idiopathic Arthritis; VKH: Vogt–Koyanagi–Harada; CMV: Cytomegalovirus; BD: Behçet's disease.

Diagnosis	Type of Uveitis	Prevalence (%)	Relative Risk of Glaucoma
JIA	anterior	26.19	2.49
VKH	panuveitis	3.96	2.71
CMV	anterior	4.76	1.46
BD	panuveitis	4.76	1.46
Idiopathic	all	30.15	1.28

3. Pathogenesis of IOP Dysregulation in Uveitic Glaucoma

The pathogenesis of IOP dysregulation in uveitis is not fully understood yet. Uveitis can compromise the delicate balance between secretion and outflow of aqueous humor; therefore, in patients with uveitis, both ocular hypotony and hypertension can occur.

It has been reported that ocular hypertension develops in up to 46% of uveitic patients and that it is more common in patients with chronic inflammation than in those with acute uveitis [32,33]. Ocular hypotony is less common in uveitis, affecting up to 10% of patients and is more common in young patients, mainly those with JIA uveitis [34]. IOP reduction is frequently due to ciliary body inflammation, resulting in decreased aqueous production [35] and can occur as a complication following glaucoma surgery.

Interestingly, raised IOP in uveitis can be documented in eyes with either closed or open iridocorneal angle. In uveitic glaucoma, open-angle is more common than angle closure [36]. Several pathogenetic mechanisms have been hypothesized to cause IOP rise in both closed- and open-angle uveitic glaucoma.

In the closed angle UG, IOP rise can be secondary to three mechanisms: (1) pupillary block caused by posterior synechiae, (2) presence of peripheral anterior synechiae, and (3) forward rotation of the ciliary body, described in VKH Syndrome [37].

In open-angle UG, the disruption of the blood-aqueous barrier due to inflammation increases the level of proteins in the aqueous humor in uveitic eyes. This great amount of proteins (up to 1308 mg vs. 30–50 mg in normal eyes) can interfere with the aqueous outflow. Peretz et al., characterized the protein pattern of the aqueous humor of patients with and without uveitis. The results were very similar to those of serum proteins, supporting the theory that the increased blood-aqueous barrier permeability may be involved in the pathogenesis of uveitis [38]. Ladas et al., demonstrated a relationship between aqueous humor protein concentration and aqueous humor outflow. Specifically, the outflow was significantly reduced in uveitis with high aqueous humor protein levels and appeared to be normal in active uveitis with low flare levels [39]. In addition, mechanical clogging of the Trabecular Meshwork (TM) can result from the high concentration of inflammatory cells and debris in the anterior chamber [38].

The inflammation of the TM, known as trabeculitis, is also supposed to be involved in the development of uveitic open-angle glaucoma. Trabeculitis is frequently characterized by an acute IOP increase, as in the Fuchs heterochromic uveitis (FHU), herpetic uveitis (HU), and Posner–Schlossman syndrome (PSS) subtypes [39].

Murray et al., collected aqueous samples during cataract surgery from 22 patients with inactive uveitis and 24 subjects without uveitis. Patients with uveitis showed a greater expression of interleukin 2 (IL-2), interferon-gamma (INF-γ), and tumor necrosis factor-alpha (TNF-α), related to T-helper 1 inflammatory response, compared to patients without uveitis [40]. Current evidence shows that the T-helper response is involved in uveitis pathogenesis, and the cytokine response may differ in infectious and non-infectious uveitis [41,42]. However, there is still debate whether a specific cytokine pattern may be linked to a higher risk of developing uveitic glaucoma.

Uveitis is an inflammatory disease, and glucocorticoids (GCs) are considered the first-line treatment. In some patients, steroid responsiveness may play a part in the pathogenesis of uveitic glaucoma. The most accepted definition of steroid responsiveness is a clinically relevant IOP increasing more than 10 mmHg over the baseline, which can lead to glaucomatous optic neuropathy over time [43].

In adults, approximately 62% do not show a significant IOP increase after GCs administration [43,44]. Still, GCs responsiveness has a significant impact in children: about 60% of the pediatric population shows a significant IOP increase [43,45] with a higher risk of Steroid-Induced Glaucoma (SIG) than adults.

GCs induce the expression of genes involved in microstructural changes in the trabecular meshwork. The overexpression of inhibitors of Metalloproteinases (MMPs) increases the deposition of extracellular matrix in the TM, resulting in increased stiffness [43,46]. In addition, GCs stimulate the rearrangement of the cytoskeletal network with the production of cross-linked actin networks (CLANs) that inhibits TM contractility [43,47–49]. Dexamethasone and prednisolone induce IOP elevation more often than other GCs, [43,49–53] and topical route administration is more frequently associated with SIG than other routes [43,54,55]. Ocular hypertension occurs in 3–6 weeks after continuous eye drops administration, and IOP usually returns to the baseline values within 2 weeks after the end of treatment [43,56]. However, IOP elevation can be irreversible after GCs cessation once microstructural damage of the trabecular meshwork occurs [43].

4. Medical Management

Correct management of the systemic disease underlying the uveitis is crucial to minimize the recurrence of intraocular inflammation and becomes pivotal in preserving visual function in these patients. The use of systemic steroids and/or immunosuppressive medication in pediatric patients requires multidisciplinary collaboration and careful follow-up to monitor possible side effects. Revising the whole medical management for the systemic disease related to uveitis is beyond the scope of this paper; Table 2 outlines the most relevant therapeutic options in the pediatric population [57].

Table 2. Medical options in pediatric uveitis.

Glucocorticoids
Methotrexate
Anti-TNF-alpha monoclonal antibody therapy (adalimumab, infliximab)
Anti-interleukin 6 receptor antibody therapy (tocilizumab)
A CTLA-4-Ig fusion protein (abatacept)
Anti-CD20 monoclonal antibody therapy (rituximab)
Janus kinase inhibitors (tofacitinib)

The medical management of UG in children is complex, and the clinician should consider the etiology, the patient's age at presentation and general health and be aware of the known efficacy and safety profiles of each drug. All the IOP-lowering agents are indicated in UG in adults, except for prostaglandin analogs during active inflammation and pilocarpine, whose miotic effect may increase the risk of posterior synechiae [58].

Nowadays, there is a lack of RCTs regarding the treatment of childhood uveitic glaucoma.

Sacchi et al. [59] identified only five RCTs [60–64] evaluating the efficacy of topical IOP-lowering drugs in patients younger than 18 with glaucoma and ocular hypertension (Table 3).

Table 3. Main RCTs for medical treatment in pediatric glaucoma.

Study	Topical IOP Lowering Drugs	Efficacy in IOP Reduction
Ott et al. (2005) [60]	Dorzolamide 2% vs. timolol gel 0.25–0.50%	20–23% vs. 25%
Whitson et al. (2008) [61]	Brinzolamide 1% vs. levobetaxolol 0.5%	20% vs. 16%
Plager et al. (2009) [62]	Betaxolol 0.25% vs. timolol gel 0.25–0.50%	9% vs. 12.15%
Maeda-Chubachi et al. (2011) [63]	Latanoprost 0.005% vs. timolol 0.25–0.50%	26% vs. 21%
Dixon et al. (2017) [64]	Travoprost with 0.5% vs. timolol 0.25–0.50%	27% vs. 25%

However, most patients were diagnosed with PCG and JOAG, and secondary glaucoma other than UG was included. Therefore, no specific conclusions for medical therapy of UG can be drawn from the primary RCTs on pediatric glaucoma.

Even if most glaucoma eye drops are not licensed for children, they currently have an off-label use in pediatric glaucoma, with a good safety profile [65].

Beta-blockers (BBs) and Carbonic Anhydrase Inhibitors (CAIs) are considered first-line options in UG [58]. Topical BBs decrease aqueous production, reducing IOP by approximately 20–25% in adults and up to 36% in children [59].

Effectiveness in IOP lowering is similar among the different beta-blockers agents. Timolol and Betaxolol are available in eye drops, in 0.25% and 0.50% concentrations, with twice daily administration. Timolol is also available in a 0.25% or 0.50% gel with a single-day indication. BBs are primarily prescribed for pediatric glaucoma, but a careful anamnesis is pivotal to prevent systemic side effects in children at risk of hypotension, bradycardia, bronchospasm, and apnoea [65]. In the previously reported RCTs, one patient developed bradycardia, and another had pneumonia as a severe side effect after timolol administration. Cardioselective B1 blockers, such as Betaxolol, are less likely to cause respiratory side effects.

CAIs, such as Dorzolamide or Brinzolamide, decrease aqueous production and reduce the IOP by approximately 20% in adults and 23% in children [59]. CAIs eye drops three times a day in monotherapy seem more effective in children than in adults [59]. Therefore, Dorzolamide or Brinzolamide could be an excellent pharmacological choice in young patients. However, a history of allergy to sulfonamides should be excluded before CAIs prescription because of the possible cross-reactivity between the two classes of drugs.

Corneal edema, superficial punctate keratitis and stinging have been reported as side effects. Systemic adverse events observed in oral administration, such as renal failure, metabolic acidosis or aplastic anemia, are uncommon in topical treatment [65].

Although Prostaglandin Analogs (PgAs) are considered the first-line treatment for non-uveitic glaucoma, their use is not recommended in uveitic eyes because of their pro-inflammatory effect. However, they can be safely used in quiescent uveitis undergoing immunomodulatory therapy. PgAs increase the uveoscleral outflow and are the most effective pharmaceutical option in monotherapy, with IOP reduction of up to 35% in adults and up to 27% in children [59]. They have the best systemic safety profile and show mainly local side effects such as ocular irritation, itchiness, iris and eyelid pigmentation, periocular fat atrophy and eyelash elongation [58]. In addition, reactivation of herpes keratitis, exacerbation of uveitis and macular edema have been reported [66,67]. Latanoprost and Travoprost administered once a day shows an excellent efficacy and safety profile in pediatric clinical trials compared to Timolol or Dorzolamide. However, the IOP-lowering effect seems to be poorer in children than in adults [59].

A phase III trial designed to assess the safety and efficacy of Bimatoprost compared to Timolol in children was prematurely discontinued in 2015 due to insufficient enrollment. (Clinicaltrial.gov identifier: NCT01068964)

Alpha Adrenergic Agonists (AAAs), such as Brimonidine tartrate, are scarcely studied in children. These drugs can activate alpha-2 receptors in the bulbar vasomotor center resulting in life-threatening side effects in patients under 23 months, such as apnoea, lethargy, hypotension, bradycardia and hypothermia. These severe adverse events may be related to an immature and permeable blood-brain barrier, so AAAs are not safely prescribed for children under 2 years old and should be used with caution in older children [65].

5. Surgical Management

Surgery is a viable therapeutic option when glaucoma progresses, regardless of maximal medical therapy. However, appropriate planning of glaucoma surgery in pediatric patients is still challenging compared to adults [68,69] because of the poor collaboration affecting the results of examinations, including visual field testing and IOP measurement.

In pediatric UG, there is an intensive wound-healing response due to young age and uveitic inflammation [70] that can affect surgical outcomes. Furthermore, in young patients, preservation of the conjunctiva is crucial for future filtration surgeries.

Unfortunately, RCTs specifically designed for UG in the pediatric population are not available, and for this reason, surgical treatment is not based on strong evidence [70].

According to available literature, goniotomy may be a good first-line treatment for pediatric UG [71,72] because it is minimally invasive, repeatable and preserves the conjunctiva. In addition, vision-threatening complications are very uncommon after this procedure.

Brenda L. et al., in a retrospective case series, reported that a single goniotomy effectively controls IOP in pediatric uveitic glaucoma, with 50% of success at 10 years; a second goniotomy can raise the rate of success up to 70% in 10 years [71].

Goniotomy requires a clear view of the angle, so eyes with narrow or closed angles, band keratopathy, or corneal edema are unsuitable for this approach. In these conditions, ab externo trabeculotomy is preferred.

Qianqian Wang et al., investigated the role of ab externo trabeculotomy in 28 UG eyes. The Authors reported an overall surgical success rate of 81.8% with one or two procedures. The survival probability was similar to those reported for goniotomy [73]. Moreover, despite the greater conjunctival manipulation, a temporal approach could preserve the superior conjunctiva for future glaucoma surgeries.

Filtration surgery, such as trabeculectomy and Glaucoma Drainage Devices (GDDs), is another option designed to drive aqueous humor from the anterior chamber to an external conjunctival filtering bleb. As previously mentioned, in pediatric UG, the intensive wound-healing response due to young age and chronic inflammation [68] may lead to subconjunctival fibrosis and surgical failure.

Antifibrotic agents, such as Mitomycin C (MMC) or 5-fluorouracil (5FU), have significantly improved the long-term rate of success of filtration surgery, but their use is off-label in ophthalmology [74,75]. Although these drugs are currently used in many pediatric fields (such as oncology, otorhinolaryngology, gastroenterology or ophthalmology) [76–78], neither MMC nor 5 FU is officially licensed for children.

Nowadays, postoperative fibrosis remains a significant problem in glaucoma management, especially in uveitic patients. GDDs' success is less related to sub-conjunctival fibrosis than trabeculectomy, so they are a good first choice in UG [69]; some models are commercially available in smaller sizes, suitable for children (Table 4).

Table 4. Main Glaucoma Drainage Devices available for pediatric patients.

Type	Model	Plate Area (mm^2)	Plate Material
Ahmed® Implant *			
Pediatric size	S3	96	polypropylene
Pediatric size	FP8	96	silicone
Pars plana (Ped)	PC8	96	silicone
Pars plana (Ped)	PS3	96	polypropylene
Molteno3 ® Implant **			
Pediatric	P1	80	polypropylene

* valved; ** non valved.

Although smaller models have been specifically designed for children, there is a lack of studies comparing the effectiveness of pediatric vs. adult models in the pediatric population. The little available literature does not seem to demonstrate the superiority of one model over the other [79,80]

Interestingly, Hye Jin Kwon et al., reported similar outcomes in GDD and trabeculectomy with anti-fibrotic agents at 5 years in 82 uveitic eyes in adults. However, trabeculectomy was more likely to fail when reactivation of uveitis occurred because of subconjunctival fibrosis [81].

Unfortunately, there is a lack of studies comparing the outcomes of trabeculectomy and GDDs in uveitic children. The surgeon should consider the high risk of trabeculectomy failure in these patients: the overall success rate at ≥5 years ranges from 16% to 73% [82].

GDDs' success rate at ≥5 years is higher, ranging from 38% to 89% [82]. Nevertheless, GDDs have considerable long-term complications, such as corneal swelling and tube erosion through the conjunctiva.

Therefore, the choice of filtration surgery should be carefully evaluated according to the clinical situation of the young patient. Pediatric uveitic patients often show an early onset of cataracts due to steroid therapy and prolonged ocular inflammation. Clear vision is pivotal to avoid amblyopia in the pediatric population; therefore, cataract extraction is often necessary for children diagnosed with uveitis to achieve acceptable visual outcomes.

Cataract surgery in UG children is challenging for different reasons. First, posterior synechiae, inflammatory membranes on the anterior capsule, small pupils scarcely responsive to mydriatics and capsular bag instability may be intraoperative pitfalls in uveitic eyes. Furthermore, postoperative complications such as macular edema and epiretinal membrane are more likely to occur after phacoemulsification, and any effort should be made to control ocular inflammation before and after cataract surgery in patients with a history of uveitis [83].

Second, the correct timing of cataract surgery in children is still debated. Multiple factors must be considered prior to surgery, such as the patient's age, risk of amblyopia, and the uni- or bilaterality of the cataract. In addition, in childhood Ugs, the timing of phacoemulsification in relation to glaucoma filtering surgery is also critical: if performed separately, phacoemulsification may increase the risk of failure of glaucoma surgery, and glaucoma surgery may accelerate cataract development [84]. Therefore, pediatric patients must be evaluated on a case-by-case basis, using extreme caution, before the final decision is made. Posterior capsulotomy and anterior vitrectomy are recommended in children, and some Authors suggest a complete vitrectomy for young patients with uveitis and cataract associated with JIA [83]. Hydrophobic acrylic Intraocular Lenses (IOLs) are considered the best option for patients affected by uveitis. However, inflammatory membranes and deposits may occur after phacoemulsification with an IOL implant, especially in patients with sarcoidosis and JIA [83]. IOL implant is not recommended in infants younger than 1 year.

Choosing the IOL power is not easy in pediatric patients because of continuous ocular growth during childhood. An under-correction of the IOL power, according to the age of the patient, with a proper refractive correction, usually achieves acceptable results [85].

Surgical aphakia could be an option in patients with uncertain control of inflammation or capsular bag instability. However, IOL implantation is associated with a lower risk of severe hypotony after surgery [86].

6. Conclusions and Special Considerations for Children

Childhood uveitic glaucoma is a potentially blinding condition. Clinical management is challenging for ophthalmologists and requires a multidisciplinary approach. Children have a long life expectancy, so a prompt diagnosis is extremely important, as well as a proper IOP-lowering therapy, in order to reduce disease progression and preserve vision for life. Unfortunately, there is a lack of evidence from RCTs in the pediatric population [59] and especially in children with uveitic glaucoma. Most of the data for UG in the pediatric population were collected from retrospective studies with small sample sizes.

In this review, we aimed to summarize the best available evidence that could be useful for ophthalmologists, pediatricians, rheumatologists and other specialists.

A crucial topic is how to define the pediatric population and how to manage different age groups. The International Conference on Harmonisation/Committee for Medicinal Products for Human Use (ICH/CHMP) guidelines define five subgroups in the pediatric population for clinical trials (Table 5).

Table 5. Pediatric population subgroups according to the International Conference on Harmonisation/Committee for Medicinal Products for Human Use (ICH/CHMP).

Pediatric Subgroup	Age
Preterm newborn infants	<37 gestation week
Term newborn infants	0–27 days
Infants and toddlers	28 days–23 months
Children	2–11 years
Adolescents	12 to 16–18 years (depending on region)

This heterogeneous definition makes it difficult to provide general results in clinical trials that could be valid for different ages.

Medical therapy is usually the first line of treatment in glaucoma, but we should remember that children are more prone to systemic side effects than adults because of their smaller body mass and blood volume and the immaturity of the brain-blood barrier [65].

Any effort to decrease the systemic absorption of topical treatments is crucial in these young patients. Training parents and caregivers to occlude the lacrimal duct for 2–3 min after eye drop instillation could be useful in long-term medical management. Furthermore, poor compliance with therapy may be more common in the pediatric population, and the resulting suboptimal IOP control increases the risk of poor visual outcomes in glaucoma.

Fixed combinations, possibly in preservative-free formulation, could be a good choice to reduce the number of daily administrations and improve tolerability and compliance. Currently, none of the fixed anti-glaucoma combinations is licensed for children [65].

Anterior chamber implants could be a new therapeutic strategy to overcome compliance issues in children, preventing ocular surface disease induced by topical therapy and preserving healthy conjunctiva for future surgeries. Durysta® (AbbVie, an Allergan company) is a Bimatoprost anterior chamber implant, the only Sustained Release (SR) approved by the Food and Drug Administration (FDA) available for glaucoma treatment in 2022. Two phase 1 studies (Clinicaltrial.gov identifiers: NCT04060758, NCT05333419) are recruiting patients to evaluate the safety, tolerability and effective dose of an SR Latanoprost implant. A phase 1 study (NCT04360174), two phase 2 studies (NCT02371746 NCT05335122), and two phase 3 studies (NCT03868124 NCT03519386) are trying to assess the efficacy of an intraocular Travoprost implant in glaucoma patients. Unfortunately, none of these new devices has been licensed for children yet, and new clinical trials are needed to assess safety and efficacy in the pediatric population.

A valid therapeutic option for UG could be Rho Kinases Inhibitors (RKIs), such as Netarsudil and Ripasudil, a new pharmacological option in glaucoma treatment. RKIs increase aqueous outflow through the trabecular meshwork, reducing IOP by about 20–25% in adults in monotherapy. Conjunctival hyperemia, conjunctival hemorrhage and cornea verticillata are the most common side effects [58]. Phase I studies show that RKIs are involved in anti-inflammatory molecular pathways, and so they might be helpful in UG management [87]. Moreover, Kusuhara et al., evaluated Ripasudil eye drops administered twice a day as monotherapy or add-on treatment in 21 eyes affected with uveitic glaucoma in a retrospective case series. The Authors reported no reactivation of uveitis, with effectiveness and safety comparable to other glaucoma types [88]. Although no data are available for children yet, in the future, RKIs might be a specific therapeutic option for uveitic glaucoma in the pediatric population.

In current opinion, surgery is considered the mainstay treatment of pediatric glaucoma for different reasons. First, according to the literature, pediatric patients have an increased probability of non-response to medical therapy [59] and of side effects. In our opinion, surgery should also be considered for children not compliant with topical therapy.

Moreover, young patients have a long life expectancies, long disease durations, and are more likely to need multiple surgeries during their life. Therefore the surgeon should choose an approach that ensures effective IOP control over years and minimizes conjunctival scarring to prevent the failure of future surgeries. In addition, pediatric patients may not tolerate postoperative discomfort, so they need careful postoperative management to avoid complications. Available literature does not suggest significant differences in effectiveness between the surgical approaches described above [80], but a stable control of the disease underlying the uveitis is strongly recommended to avoid further reactivation and possible surgical failure [57].

Nowadays, the surgical treatment of pediatric UG has scarce evidence-based guidelines. Most authors strive to collect large datasets for clinical studies, and reliable prospective analysis is often impossible. Therefore the results are unpowered compared to other pediatric glaucoma types. To provide further insight into this particular topic in a fragile population, large, controlled, well-designed, and prospective studies with longer follow-ups are needed.

In conclusion, pediatric UG is a complex disease requiring multidisciplinary management and careful follow-up. Ophthalmologists, pediatricians, and rheumatologists have to work in close tandem to achieve the highest quality of life for patients and their caregivers [89].

Author Contributions: Conceptualization: V.I. and P.M.; investigation: V.I., P.M. and G.M.; methodology, A.B. and A.L.; writing—original draft preparation, V.I., P.M. and G.M.; writing—review and editing, S.G., F.G. and A.B.; project administration: A.B.; supervision, A.L. All authors have read and agreed to the published version of the manuscript.

Funding: This research received no external funding.

Institutional Review Board Statement: Not applicable.

Informed Consent Statement: Not applicable.

Data Availability Statement: All data analyzed during this study are included in this published article.

Conflicts of Interest: The authors declare no conflict of interest.

References

1. Marchini, G.; Toscani, M.; Chemello, F. Pediatric glaucoma: Current perspectives. *Pediatr. Health Med. Ther.* **2014**, *5*, 15–27. [CrossRef]
2. Gilbert, C.; Foster, A. Childhood blindness in the context of VISION 2020—The right to sight. *Bull. World Health Organ.* **2001**, *79*, 227–232.
3. Thau, A.; Lloyd, M.; Freedman, S.; Beck, A.; Grajewski, A.; Levin, A.V. New classification system for pediatric glaucoma: Implications for clinical care and a research registry. *Curr. Opin. Ophthalmol.* **2018**, *29*, 385–394. [CrossRef]
4. Gautam Seth, N.; Yangzes, S.; Thattaruthody, F.; Singh, R.; Bansal, R.; Raj, S.; Kaushik, S.; Gupta, V.; Pandav, S.S.; Ram, J.; et al. Glaucoma Secondary to Uveitis in Children in a Tertiary Care Referral Center. *Ocul. Immunol. Inflamm.* **2019**, *27*, 456–464. [CrossRef]
5. Heinz, C.; Koch, J.M.; Zurek-Imhoff, B.; Heiligenhaus, A. Prevalence of uveitic secondary glaucoma and success of nonsurgical treatment in adults and children in a tertiary referral center. *Ocul. Immunol. Inflamm.* **2009**, *17*, 243–248. [CrossRef]
6. Holland, G.N.; Stiehm, E.R. Special considerations in the evaluation and management of uveitis in children. *Am. J. Ophthalmol.* **2003**, *135*, 867–878. [CrossRef]
7. Kanski, J.J. Uveitis in juvenile chronic arthritis: Incidence, clinical features and prognosis. *Eye* **1988**, *2*, 641–645. [CrossRef]
8. Kaur, S.; Kaushik, S.; Singh Pandav, S. Pediatric Uveitic Glaucoma. *J. Curr Glaucoma Pract.* **2013**, *7*, 115–117. [CrossRef]
9. Papadopoulos, M.; Cable, N.; Rahi, J.; Khaw, P.T.; BIG Eye Study Investigators. The British Infantile and Childhood Glaucoma (BIG) Eye Study. *Investig. Ophthalmol. Vis. Sci.* **2007**, *48*, 4100–4106. [CrossRef]
10. Paroli, M.P.; Speranza, S.; Marino, M.; Pirraglia, M.P.; Pivetti-Pezzi, P. Prognosis of juvenile rheumatoid arthritis-associated uveitis. *Eur. J. Ophthalmol.* **2003**, *13*, 616–621. [CrossRef]
11. De Boer, J.; Wulffraat, N.; Rothova, A. Visual loss in uveitis of childhood. *Br. J. Ophthalmol.* **2003**, *87*, 879–884. [CrossRef]
12. Jabs, D.A.; Nussenblatt, R.B.; Rosenbaum, J.T.; Standardization of Uveitis Nomenclature (SUN) Working Group. Standardization of uveitis nomenclature for reporting clinical data. Results of the First International Workshop. *Am. J. Ophthalmol.* **2005**, *140*, 509–516. [CrossRef]
13. Petty, R.E.; Southwood, T.R.; Manners, P.; Baum, J.; Glass, D.N.; Goldenberg, J.; He, X.; Maldonado-Cocco, J.; Orozco-Alcala, J.; Prieur, A.-M.; et al. International League of Associations for Rheumatology classification of juvenile idiopathic arthritis: Second revision, Edmonton, 2001. *J. Rheumatol.* **2004**, *31*, 390–392.
14. Bullock, J.; Rizvi, S.A.A.; Saleh, A.M.; Ahmed, S.S.; Do, D.P.; Ansari, R.A.; Ahmed, J. Rheumatoid Arthritis: A Brief Overview of the Treatment. *Med. Princ. Pract.* **2018**, *27*, 501–507. [CrossRef]
15. Heiligenhaus, A.; Heinz, C.; Edelsten, C.; Kotaniemi, K.; Minden, K. Review for disease of the year: Epidemiology of juvenile idiopathic arthritis and its associated uveitis: The probable risk factors. *Ocul. Immunol. Inflamm.* **2013**, *21*, 180–191. [CrossRef]
16. Carvounis, P.E.; Herman, D.C.; Cha, S.; Burke, J.P. Incidence and outcomes of uveitis in juvenile rheumatoid arthritis, a synthesis of the literature. *Graefe's Arch. Clin. Exp. Ophthalmol.* **2006**, *244*, 281–290. [CrossRef]
17. Paroli, M.P.; Del Giudice, E.; Giovannetti, F.; Caccavale, R.; Paroli, M. Management Strategies of Juvenile Idiopathic Arthritis-Associated Chronic Anterior Uveitis: Current Perspectives. *Clin. Ophthalmol.* **2022**, *16*, 1665–1673. [CrossRef]
18. Madigan, W.P.; Raymond, W.R.; Wroblewski, K.J.; Thebpatiphat, N.; Birdsong, R.H.; Jaafar, M.S. A review of pediatric uveitis: Part, I. Infectious causes and the masquerade syndromes. *J. Pediatr. Ophthalmol. Strabismus* **2008**, *45*, 140–149. [CrossRef]
19. Nikkhah, H.; Ramezani, A.; Ahmadieh, H.; Soheilian, M.; Azarmina, M.; Dehghan, M.-H.; Moradian, S.; Nourinia, R. Childhood pars planitis; clinical features and outcomes. *J. Ophthalmic Vis. Res.* **2011**, *6*, 249–254.
20. Ruet, A. Update on pediatric-onset multiple sclerosis. *Rev. Neurol.* **2018**, *174*, 398–407. [CrossRef]
21. Ozdal, P.C.; Berker, N.; Tugal-Tutkun, I. Pars Planitis: Epidemiology, Clinical Characteristics, Management and Visual Prognosis. *J. Ophthalmic. Vis. Res.* **2015**, *10*, 469–480. [CrossRef]
22. An, Q.; Fan, C.H.; Xu, S.M. Childhood multiple sclerosis: Clinical features and recent developments on treatment choices and outcomes. *Eur Rev. Med. Pharmacol. Sci.* **2018**, *22*, 5747–5754. [CrossRef]
23. Sijssens, K.M.; Rothova, A.; Berendschot, T.T.; de Boer, J.H. Ocular hypertension and secondary glaucoma in children with uveitis. *Ophthalmology* **2006**, *113*, 853–859.e2. [CrossRef]
24. Giles, C.L. Uveitis in childhood—Part III. *Posterior. Ann. Ophthalmol.* **1989**, *21*, 23–28.
25. De Jong, P.T. Ocular toxoplasmosis; common and rare symptoms and signs. *Int. Ophthalmol.* **1989**, *13*, 391–397. [CrossRef]
26. Accorinti, M.; Bruscolini, A.; Pirraglia, M.P.; Liverani, M.; Caggiano, C. Toxoplasmic retinochoroiditis in an Italian referral center. *Eur. J. Ophthalmol.* **2009**, *19*, 824–830. [CrossRef]
27. Rao, N.A.; Gupta, A.; Dustin, L.; Chee, S.P.; Okada, A.A.; Khairallah, M.; Bodaghi, B.; Lehoang, P.; Accorinti, M.; Mochizuki, M.; et al. Frequency of distinguishing clinical features in Vogt-Koyanagi-Harada disease. *Ophthalmology* **2010**, *117*, 591–599.e1. [CrossRef]
28. Accorinti, M.; Saturno, M.C.; Iannetti, L.; Manni, P.; Mastromarino, D.; Pirraglia, M.P. Treatment and Prognosis of Vogt-Koyanagi-Harada Disease: Real-Life Experience in Long-Term Follow-Up. *J. Clin. Med.* **2022**, *11*, 3632. [CrossRef]
29. Rathinam, S.R.; Vijayalakshmi, P.; Namperumalsamy, P.; Nozik, R.A.; Cunningham, E.T., Jr. Vogt-Koyanagi-Harada syndrome in children. *Ocul. Immunol. Inflamm.* **1998**, *6*, 155–161. [CrossRef]

30. Kaza, H.; Tyagi, M.; Agarwal, K.; Behera, S.; Pappuru, R.R.; Mohan, S.; Saldanha, M.; Videkar, C.; Basu, S.; Pathengay, A.; et al. Vogt Koyanagi Harada Disease in Paediatric Age Group: Clinical Characteristics, Remission, Recurrences and Complications in Asian Indian Population. *Semin. Ophthalmol.* **2022**, *37*, 187–192. [CrossRef]
31. Kesen, M.R.; Goldstein, D.A.; Tessler, H.H. Uveitis associated with pediatric behçet disease in the american midwest. *Am. J. Ophthalmol.* **2008**, *146*, 819–827.e2. [CrossRef]
32. Kanda, T.; Shibata, M.; Taguchi, M.; Ishikawa, S.; Harimoto, K.; Takeuchi, M. Prevalence and aetiology of ocular hypertension in acute and chronic uveitis. *Br. J. Ophthalmol.* **2014**, *98*, 932–936. [CrossRef]
33. Herbert, H.M.; Viswanathan, A.; Jackson, H.; Lightman, S.L. Risk factors for elevated intraocular pressure in uveitis. *J. Glaucoma* **2004**, *13*, 96–99. [CrossRef]
34. Sen, H.N.; Drye, L.T.; Goldstein, D.A.; Larson, T.A.; Merrill, P.; Pavan, P.R.; Sheppard, J.D.; Burke, A.; Srivastava, S.K.; Jabs, D.A.; et al. Hypotony in patients with uveitis: The Multicenter Uveitis Steroid Treatment (MUST) Trial. *Ocul. Immunol. Inflamm.* **2012**, *20*, 104–112. [CrossRef]
35. Agrawal, R.V.; Murthy, S.; Sangwan, V.; Biswas, J. Current approach in diagnosis and management of anterior uveitis. *Indian J. Ophthalmol.* **2010**, *58*, 11–19. [CrossRef]
36. Kalogeropoulos, D.; Sung, V.C. Pathogenesis of Uveitic Glaucoma. *J. Curr. Glaucoma Pract.* **2018**, *12*, 125–138. [CrossRef]
37. Kishi, A.; Nao-i, N.; Sawada, A. Ultrasound biomicroscopic findings of acute angle-closure glaucoma in Vogt-Koyanagi-Harada syndrome. *Am. J. Ophthalmol.* **1996**, *122*, 735–737. [CrossRef]
38. Peretz, W.L.; Tomasi, T. Aqueous humor proteins in uveitis. Immunoelectrophoretic and gel diffusion studies on normal and pathological human aqueous humor. *Arch. Ophthalmol.* **1961**, *65*, 20–23. [CrossRef]
39. Ladas, J.G.; Yu, F.; Loo, R.; Davis, J.L.; Coleman, A.L.; Levinson, R.D.; Holland, G.N. Relationship between aqueous humor protein level and outflow facility in patients with uveitis. *Investig. Ophthalmol. Vis. Sci.* **2001**, *42*, 2584–2588.
40. Murray, P.I.; Clay, C.D.; Mappin, C.; Salmon, M. Molecular analysis of resolving immune responses in uveitis. *Clin. Exp. Immunol.* **1999**, *117*, 455–461. [CrossRef]
41. Murphy, C.C.; Duncan, L.; Forrester, J.V.; Dick, A.D. Systemic CD4(+) T cell phenotype and activation status in intermediate uveitis. *Br. J. Ophthalmol.* **2004**, *88*, 412–416. [CrossRef]
42. Deschênes, J.; Char, D.H.; Kaleta, S. Activated T lymphocytes in uveitis. *Br. J. Ophthalmol.* **1988**, *72*, 83–87. [CrossRef]
43. Roberti, G.; Oddone, F.; Agnifili, L.; Katsanos, A.; Michelessi, M.; Mastropasqua, L.; Quaranta, L.; Riva, I.; Tanga, L.; Manni, G. Steroid-induced glaucoma: Epidemiology, pathophysiology, and clinical management. *Surv. Ophthalmol.* **2020**, *65*, 458–472. [CrossRef]
44. Armaly, M.F.; Becker, B. Intraocular pressure response to topical corticosteroids. *Fed. Proc.* **1965**, *24*, 1274–1278.
45. Ng, J.S.; Fan, D.S.; Young, A.L.; Yip, N.K.; Tam, K.; Kwok, A.K.; Lam, D.S. Ocular hypertensive response to topical dexamethasone in children: A dose-dependent phenomenon. *Ophthalmology* **2000**, *107*, 2097–2100. [CrossRef]
46. Samples, J.R.; Alexander, J.P.; Acott, T.S. Regulation of the levels of human trabecular matrix metalloproteinases and inhibitors by interleukin-1 and dexamethasone. *Investig. Ophthalmol. Vis. Sci.* **1993**, *34*, 3386–3395.
47. Clark, A.F.; Brotchie, D.; Read, A.T.; Hellberg, P.; English-Wright, S.; Pang, I.H.; Ethier, C.R.; Grierson, I. Dexamethasone alters F-actin architecture and promotes cross-linked actin network formation in human trabecular meshwork tissue. *Cell Motil. Cytoskelet.* **2005**, *60*, 83–95. [CrossRef]
48. Clark, A.F.; Miggans, S.T.; Wilson, K.; Browder, S.; McCartney, M.D. Cytoskeletal changes in cultured human glaucoma trabecular meshwork cells. *J. Glaucoma* **1995**, *4*, 183–188.
49. Clark, A.F.; Wilson, K.; McCartney, M.D.; Miggans, S.T.; Kunkle, M.; Howe, W. Glucocorticoid-induced formation of cross-linked actin networks in cultured human trabecular meshwork cells. *Investig. Ophthalmol. Vis. Sci.* **1994**, *35*, 281–294.
50. Abadia, B.; Calvo, P.; Ferreras, A.; Bartol, F.; Verdes, G.; Pablo, L. Clinical Applications of Dexamethasone for Aged Eyes. *Drugs Aging* **2016**, *33*, 639–646. [CrossRef]
51. Cantrill, H.L.; Palmberg, P.F.; Zink, H.A.; Waltman, S.R.; Podos, S.M.; Becker, B. Comparison of in vitro potency of corticosteroids with ability to raise intraocular pressure. *Am. J. Ophthalmol.* **1975**, *79*, 1012–1017. [CrossRef]
52. Laurell, C.G.; Zetterström, C. Effects of dexamethasone, diclofenac, or placebo on the inflammatory response after cataract surgery. *Br. J. Ophthalmol.* **2002**, *86*, 1380–1384. [CrossRef]
53. Ramsell, T.G.; Trillwood, W.; Draper, G. Effects of prednisolone eye drops. A trial of the effects of prednisolone phosphate eye drops on the intra-ocular pressure of normal volunteers. *Br. J. Ophthalmol.* **1967**, *51*, 398–402. [CrossRef]
54. Sheppard, J.D.; Comstock, T.L.; Cavet, M.E. Impact of the Topical Ophthalmic Corticosteroid Loteprednol Etabonate on Intraocular Pressure. *Adv. Ther.* **2016**, *33*, 532–552. [CrossRef]
55. Sihota, R.; Konkal, V.L.; Dada, T.; Agarwal, H.C.; Singh, R. Prospective, long-term evaluation of steroid-induced glaucoma. *Eye* **2008**, *22*, 26–30. [CrossRef]
56. Armaly, M.F. Effect of corticosteroids on intraocular pressure fluid dynamics II. The effect of dexamethasone in the glaucomatous eye. *Arch. Ophthalmol.* **1963**, *70*, 492–499. [CrossRef]
57. Chang, M.H.; Shantha, J.G.; Fondriest, J.J.; Lo, M.S.; Angeles-Han, S.T. Uveitis in Children and Adolescents. *Rheum. Dis. Clin. N. Am.* **2021**, *47*, 619–641. [CrossRef]
58. Spaeth, G.L.; Bagnasco, L.; Bagnis, A.; Breda, J.B.; Bonzano, C.; Brezhnev, A.; Bron, A.; Cutolo, C.A.; Cvenkel, B.; Gandolfi, S. European Glaucoma Society Terminology and Guidelines for Glaucoma, 5th edition. *Br. J. Ophthalmol.* **2021**, *105*, 1–169. [CrossRef]

59. Sacchi, M.; Lizzio, R.A.U.; Villani, E.; Monsellato, G.; Lucentini, S.; Cremonesi, E.; Luccarelli, S.; Serafino, M.; Nucci, P. Medical management of pediatric glaucoma: Lessons learned from randomized clinical trials. *Graefes Arch. Clin. Exp. Ophthalmol.* **2020**, *258*, 1579–1586. [CrossRef]
60. Ott, E.Z.; Mills, M.D.; Arango, S.; Getson, A.J.; Assaid, C.A.; Adamsons, I.A. A randomized trial assessing dorzolamide in patients with glaucoma who are younger than 6 years. *Arch. Ophthalmol.* **2005**, *123*, 1177–1186. [CrossRef]
61. Whitson, J.T.; Roarty, J.D.; Vijaya, L.; Robin, A.L.; Gross, R.D.; Landry, T.A.; Dickerson, J.E.; Scheib, S.A.; Scott, H.; Hua, S.Y.; et al. Efficacy of brinzolamide and levobetaxolol in pediatric glaucomas: A randomized clinical trial. *J. Am. Assoc. Pediatr. Ophthalmol. Strabismus* **2008**, *12*, 239–246.e3. [CrossRef]
62. Plager, D.A.; Whitson, J.T.; Netland, P.A.; Vijaya, L.; Sathyan, P.; Sood, D.; Krishnadas, S.; Robin, A.L.; Gross, R.D.; Scheib, S.A.; et al. Betaxolol hydrochloride ophthalmic suspension 0.25% and timolol gel-forming solution 0.25% and 0.5% in pediatric glaucoma: A randomized clinical trial. *J. Am. Assoc. Pediatr. Ophthalmol. Strabismus.* **2009**, *13*, 384–390. [CrossRef]
63. Maeda-Chubachi, T.; Chi-Burris, K.; Simons, B.D.; Freedman, S.F.; Khaw, P.T.; Wirostko, B.; Yan, E.; A6111137 Study Group. Comparison of latanoprost and timolol in pediatric glaucoma: A phase 3, 12-week, randomized, double-masked multicenter study. *Ophthalmology* **2011**, *118*, 2014–2021. [CrossRef]
64. Dixon, E.R.; Landry, T.; Venkataraman, S.; Gustafson, N.; Salem, C.; Bradfield, Y.; Aljasim, L.A.; Feldman, R. A 3-month safety and efficacy study of travoprost 0.004% ophthalmic solution compared with timolol in pediatric patients with glaucoma or ocular hypertension. *J. Am. Assoc. Pediatr. Ophthalmol. Strabismus* **2017**, *21*, 370–374.e1. [CrossRef]
65. Samant, M.; Medsinge, A.; Nischal, K.K. Pediatric Glaucoma: Pharmacotherapeutic Options. *Paediatr. Drugs.* **2016**, *18*, 209–219. [CrossRef]
66. Fortuna, E.; Cervantes-Castañeda, R.A.; Bhat, P.; Doctor, P.; Foster, C.S. Flare-up rates with bimatoprost therapy in uveitic glaucoma. *Am. J. Ophthalmol.* **2008**, *146*, 876–882, Correction in *Am. J. Ophthalmol.* **2009**, *147*, 565. [CrossRef]
67. Chang, J.H.; McCluskey, P.; Missotten, T.; Ferrante, P.; Jalaludin, B.; Lightman, S. Use of ocular hypotensive prostaglandin analogues in patients with uveitis: Does their use increase anterior uveitis and cystoid macular oedema? *Br. J. Ophthalmol.* **2008**, *92*, 916–921. [CrossRef]
68. Beauchamp, G.R.; Parks, M.M. Filtering surgery in children: Barriers to success. *Ophthalmology* **1979**, *86*, 170–180. [CrossRef]
69. Ramdas, W.D.; Pals, J.; Rothova, A.; Wolfs, R.C.W. Efficacy of glaucoma drainage devices in uveitic glaucoma and a meta-analysis of the literature. *Graefes Arch. Clin. Exp. Ophthalmol.* **2019**, *257*, 143–151. [CrossRef]
70. Lingham, G.; Thakur, S.; Safi, S.; Gordon, I.; Evans, J.R.; Keel, S. A systematic review of clinical practice guidelines for childhood glaucoma. *BMJ Open Ophthalmol.* **2022**, *7*, e000933. [CrossRef]
71. Bohnsack, B.L.; Freedman, S.F. Surgical outcomes in childhood uveitic glaucoma. *Am. J. Ophthalmol.* **2013**, *155*, 134–142. [CrossRef]
72. Ho, C.L.; Wong, E.Y.; Walton, D.S. Goniosurgery for glaucoma complicating chronic childhood uveitis. *Arch. Ophthalmol.* **2004**, *122*, 838–844. [CrossRef]
73. Wang, Q.; Wang, J.; Fortin, E.; Hamel, P. Trabeculotomy in the Treatment of Pediatric Uveitic Glaucoma. *J. Glaucoma* **2016**, *25*, 744–749. [CrossRef]
74. Wilkins, M.; Indar, A.; Wormald, R. Intra-operative mitomycin C for glaucoma surgery. *Cochrane Database Syst. Rev.* **2005**, *19*, CD002897. [CrossRef]
75. Cabourne, E.; Clarke, J.C.; Schlottmann, P.G.; Evans, J.R. Mitomycin C versus 5-Fluorouracil for wound healing in glaucoma surgery. *Cochrane Database Syst Rev.* **2015**, *6*, CD006259. [CrossRef]
76. Watanabe, K. Current chemotherapeutic approaches for hepatoblastoma. *Int. J. Clin. Oncol.* **2013**, *18*, 955–961. [CrossRef]
77. Senders, C.W. Use of mitomycin C in the pediatric airway. *Curr. Opin. Otolaryngol. Head Neck Surg.* **2004**, *12*, 473–475. [CrossRef]
78. Berger, M.; Ure, B.; Lacher, M. Mitomycin C in the therapy of recurrent esophageal strictures: Hype or hope? *Eur. J. Pediatr Surg.* **2012**, *22*, 109–116. [CrossRef]
79. Esporcatte, B.; Netto, C.; Tanno, T.; Melo, L.A.; Tavares, I.; de Moura, C.R. glaucoma valve FP7 and FP8 in pediatric glaucoma: A randomized clinical trial. *Investig. Ophthalmol. Vis. Sci.* **2013**, *54*, 4487.
80. Senthil, S.; Gollakota, S.; Ali, M.H.; Turaga, K.; Badakere, S.; Krishnamurthy, R.; Garudadri, C.S. Comparison of the New Low-Cost Nonvalved Glaucoma Drainage Device with Ahmed Glaucoma Valve in Refractory Pediatric Glaucoma in Indian Eyes. *Ophthalmol. Glaucoma.* **2018**, *1*, 167–174. [CrossRef]
81. Kwon, H.J.; Kong, Y.X.G.; Tao, L.W.; Lim, L.; Martin, K.R.; Green, C.; Ruddle, J.; Crowston, J.G. Surgical outcomes of trabeculectomy and glaucoma drainage implant for uveitic glaucoma and relationship with uveitis activity. *Clin. Exp. Ophthalmol.* **2017**, *45*, 472–480. [CrossRef]
82. Van Meerwijk, C.L.L.I.; Jansonius, N.M.; Los, L.I. Uveitic glaucoma in children: A systematic review on surgical outcomes. *J. Ophthalmic Inflamm. Infect.* **2022**, *12*, 35. [CrossRef]
83. Foster, C.S.; Rashid, S. Management of coincident cataract and uveitis. *Curr. Opin. Ophthalmol.* **2003**, *14*, 1–6, Correction in *Curr. Opin. Ophthalmol.* **2004**, *15*, 155. [CrossRef]
84. Kyari, F. Managing cataract surgery in patients with glaucoma. *Community Eye Health.* **2019**, *31*, 88–90.
85. Mohammadpour, M.; Shaabani, A.; Sahraian, A.; Momenaei, B.; Tayebi, F.; Bayat, R.; Mirshahi, R. Updates on management of pediatric cataract. *J. Curr. Ophthalmol.* **2018**, *31*, 118–126. [CrossRef]

86. Schmidt, D.C.; Al-Bakri, M.; Rasul, A.; Bangsgaard, R.; Subhi, Y.; Bach-Holm, D.; Kessel, L. Cataract Surgery with or without Intraocular Lens Implantation in Pediatric Uveitis: A Systematic Review with Meta-Analyses. *J. Ophthalmol.* **2021**, *2021*, 5481609. [CrossRef]
87. Uchida, T.; Honjo, M.; Yamagishi, R.; Aihara, M. The Anti-Inflammatory Effect of Ripasudil (K-115), a Rho Kinase (ROCK) Inhibitor, on Endotoxin-Induced Uveitis in Rats. *Investig. Ophthalmol. Vis. Sci.* **2017**, *58*, 5584–5593. [CrossRef]
88. Kusuhara, S.; Katsuyama, A.; Matsumiya, W.; Nakamura, M. Efficacy and safety of ripasudil, a Rho-associated kinase inhibitor, in eyes with uveitic glaucoma. *Graefes Arch. Clin. Exp. Ophthalmol.* **2018**, *256*, 809–814. [CrossRef]
89. Basilious, A.; Villani, S.; Jang, H.; Kaberi, K.M.; Malvankar-Mehta, M.S. Quality of life and caregiver burden in pediatric glaucoma: A systematic review. *PLoS ONE* **2022**, *17*, e0276881. [CrossRef]

applied sciences

MDPI

Review

The Impact of Blenderized Tube Feeding on Gastrointestinal Symptoms, a Scoping Review

Elisabetta Sforza [1,†], Domenico Limongelli [1,†], Valentina Giorgio [1,2,3], Gaia Margiotta [1], Francesco Proli [3], Eliza Maria Kuczynska [2], Chiara Leoni [2,3], Donato Rigante [1,2,3], Ilaria Contaldo [4], Chiara Veredice [4], Emanuele Rinninella [1,5], Antonio Gasbarrini [1,6], Giuseppe Zampino [1,2,3,‡] and Roberta Onesimo [2,3,*,‡]

1 Università Cattolica Sacro Cuore, 00168 Rome, Italy
2 Center for Rare Diseases and Birth Defects, Department of Woman and Child Health and Public Health, Fondazione Policlinico Universitario Agostino Gemelli IRCCS, 00168 Rome, Italy
3 Pediatric Unit, Department of Woman and Child Health and Public Health, Fondazione Policlinico Universitario Agostino Gemelli IRCCS, 00168 Rome, Italy
4 Pediatric Neuropsychiatry, Department of Woman and Child Health and Public Health, Fondazione Policlinico Universitario Agostino Gemelli IRCCS, 00168 Rome, Italy
5 Clinical Nutrition Unit, Department of Medical and Abdominal Surgery and Endocrine-Metabolic Scienze, Fondazione Policlinico Universitario Agostino Gemelli IRCCS, 00168 Rome, Italy
6 Center for Diagnosis and Treatment of Digestive Diseases, CEMAD, Gastroenterology Department, Fondazione Policlinico Universitario Agostino Gemelli IRCCS, 00168 Rome, Italy
* Correspondence: roberta.onesimo@policlinicogemelli.it
† Both authors are first authors.
‡ Both authors are last authors.

Abstract: Severe gastrointestinal symptoms are one of the main reasons for switching from conventional artificial tube feeding to blenderized tube feeding (BTF). This study aimed to describe and quantify the impact of BTF on gastrointestinal symptoms in children and adults. We analyzed four databases (PubMed, Scopus, Cochrane Library, and Google Scholar). The review was performed following the PRISMA extension for Scoping Reviews checklist. The methodological quality of articles was assessed following the NIH quality assessment tools. The initial search yielded 535 articles and, after removing duplicates and off-topic articles, 12 met the inclusion criteria. All included papers unanimously converged in defining an improvement of gastrointestinal symptoms during blenderized feeding: the eight studies involving pediatric cohorts report a decrease from 30 to over 50% in gagging and retching after commencing BTF. Similar rates are reported for constipation and diarrhea improvement in most critically ill adults. Experimental studies and particularly randomized controlled trials are needed to develop robust evidence on the effectiveness of BTF in gastrointestinal symptom improvement with prolonged follow-up and adequate medical monitoring.

Keywords: blenderized tube feeding; enteral nutrition; gastrostomy; children; disabled; review; personalized medicine

Citation: Sforza, E.; Limongelli, D.; Giorgio, V.; Margiotta, G.; Proli, F.; Kuczynska, E.M.; Leoni, C.; Rigante, D.; Contaldo, I.; Veredice, C.; et al. The Impact of Blenderized Tube Feeding on Gastrointestinal Symptoms, a Scoping Review. *Appl. Sci.* **2023**, *13*, 2173. https://doi.org/10.3390/app13042173

Academic Editors: Massimo Ralli, Gaspare Palaia and Alice Bruscolini

Received: 30 December 2022
Revised: 2 February 2023
Accepted: 6 February 2023
Published: 8 February 2023

1. Introduction

Since its introduction in the 1970s, home enteral nutrition (HEN) has been established as a reliable and effective nutritional intervention [1]. Currently, the method of choice for medium- and long-term enteral feeding is the gastrostomy tube (G-tube) [2,3], with a wide range of diets and nutrient preparations suitable for tube feeding [1]. In this context, the European Society for Clinical Nutrition and Metabolism (ESPEN) through the guidelines on HEN recently recommended using standard commercial formulas for enteral tube feeds with the exception of some specific conditions in which blended tube feeds are considered to be the first choice [1]. Specifically, standard commercial formulas refer to standard tube feed made of powdered raw materials [4]. Notably, almost all preparations of tube feedings available on the European market use nutrient isolates (except vegetable oils) and concentrates in powder instead of natural foods. The nutrients and food isolates (e.g., milk protein) are extracted from foods in these tube feedings but are provided without the natural food matrix [5].

On the contrary, BTF, also referred to as "blenderized formula" or "homemade blended formula" and "pureed by g-tube," consists of whole foods provided through a feeding tube [6,7]. Products based on real foods, such as milk, meat, and vegetables, are also commercially available for enteral tube feeding (ETF) [8]. The number of patients receiving long-term enteral nutrition along with the use of BTF has surged over the last decade. As per the Oley Foundation members' survey in 2017, most of the 216 participants, specifically pediatric (89%) and adult (66%) patients, were consuming BTF partly or totally for their nutritional needs [9]. Although guidelines currently do not recommend BTF as a first choice for tube feeding [1], many families still choose BTF over commercial formulas for several reasons, including severe gastrointestinal (GI) symptoms, intolerance to polymeric enteral formulas, or food allergies and intolerances [10]. Parenteral nutrition can play a role in relieving gastrointestinal signs/symptoms especially in children with severe neurological impairment [11]. However, given its risks and its potential to become inappropriately life sustaining, clinicians need to consider changes in conventional enteral nutrition, including blenderized formula.

This study aims to describe and quantify the impact of BTF on GI symptoms of patients without any age or diagnosis limitations.

2. Methods

2.1. Search Strategy

Supervised by R.O., E.S. performed a systematic literature search of the following databases: PubMed, Scopus, Cochrane Library, and Google Scholar. Search terms combined text words and Medical Subject Headings (MeSH), as shown in the Supplementary Table S1. Search terms included two components: terms referring to enteral tube feeding and blenderized feeding. No date limit was set. The literature search was conducted in Italy.

2.2. Study Eligibility

Following the Preferred Reporting Items for Systematic Reviews and Meta-Analyses Extension for Scoping Reviews (PRISMA-ScR) checklist [12] represented in the Supplementary Table S2, and after removing duplicates, all full-text articles were screened by two independent researchers; any discrepancies were solved in a consensus meeting. The articles were included if (a) they reported GI symptoms in pediatric and adult populations fed via blenderized tubes, (b) were freely available, and (c) written in English. Articles only assessing the nutritional value of formulas and review articles were disregarded.

2.3. Data Collection and Assessment

Data were extracted independently from included studies by two authors (E.S. and D.L.) according to a predefined data extraction sheet. Probable disagreements were solved by open discussion between the two authors, and consultation was conducted with a third author (R.O.).

Included studies were assessed independently by two researchers (E.S. and D.L.). To assess the methodological quality of all included articles, we chose three versions of NIH quality assessment tools according to study types. Due to its versatility and completeness, the NIH tool has been increasingly used in the last years for the quality assessment of articles for systematic reviews and meta-analyses. Specifically, we used the Cross-Sectional Study/Observational Cohort/Cross-Sectional Studies tool, Before-After (Pre-Post) Studies With No Control Group Studies tool, and the Controlled Intervention Studies tool [13,14]. After answering each item, two researchers rated the studies' overall quality as having a low-risk bias (good quality), moderate-risk bias (fair quality), or high-risk bias (poor quality).

3. Results

3.1. Studies Included

The initial literature search yielded 535 potentially relevant articles. After removing duplicates (n = 150), 340 "full-text" manuscripts were retrieved. Of them, 12 studies met the inclusion criteria, as shown in Figure 1. The included articles range from 2011 to 2022, spanning an 11-year period.

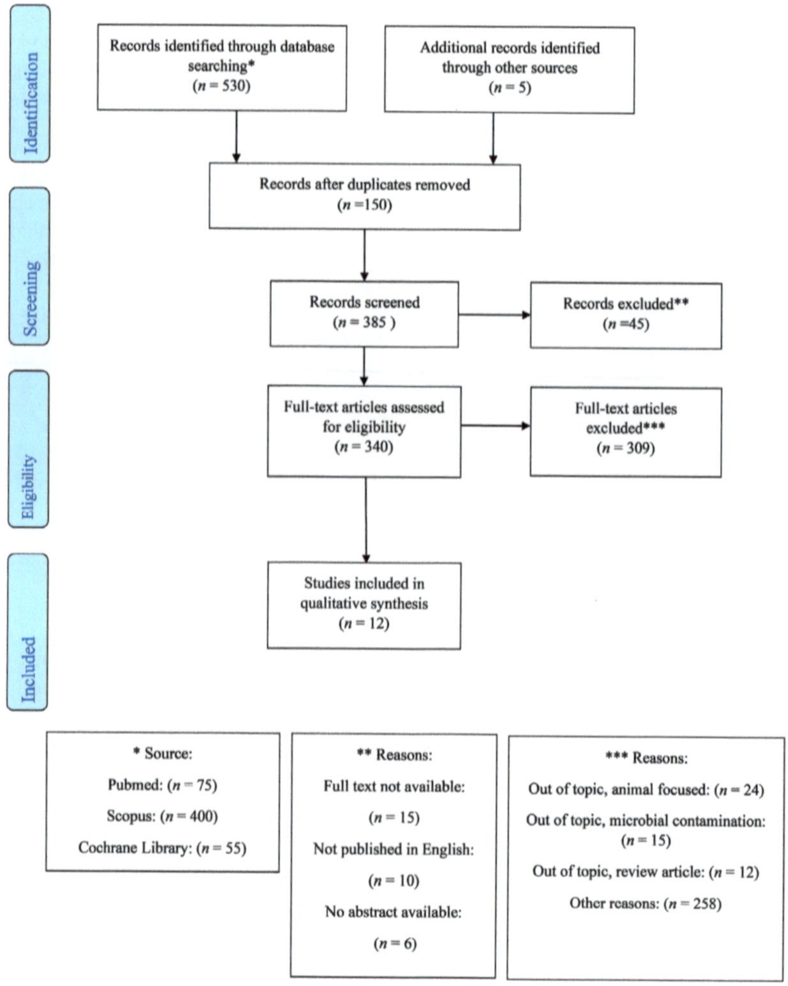

Figure 1. PRISMA diagram for article selection and inclusion in the review.

Regarding the study design, of the total studies, 3 were non-experimental cross-sectional studies (surveys) [15–17], 4 were observational retrospective studies [18–21], 5 were semi-experimental longitudinal studies [22–25], and there was only 1 RCT [5]. One study received a financial support from the private sector, specifically from Real Food Blends [25].

3.2. Assessment

By focusing on the concepts underlying the questions in the quality assessment tool, no one reached a good overall quality rating among the three cross-sectional studies. Specifically, in studies by Johnson et al., Trollip et al., and Hurt et al. [15–17], potential confounding variables were not measured and statistically adjusted for their impact on the relationship between exposure and outcome. The included observational retrospective studies [18–21] had the lowest risk of bias, although there was a lack of blindness for outcome assessors to the exposure status of participants Table 1. All the semi-experimental studies [22–25] were susceptible to some bias, reducing the quality of the results concerning sample size, follow-up loss rate, blinding of outcome assessors, and statistical analysis (Table 2). Schmidt et al. RCT [5] showed good overall quality. To note, the overall drop-out rate of participants allocated into the intervention and control groups from the study at the endpoint was high (44 to 51%, respectively). Appropriate blinding did not occur as it was not feasible (Tables 3 and 4).

Table 1. Quality Assessment of Included Studies using NIH Quality Assessment Tool for Observational Cohort and Cross-Sectional * Studies.

Study	Q1	Q2	Q3	Q4	Q5	Q6	Q7	Q8	Q9	Q10	Q11	Q12	Q13	Q14	Quality (Total Quality Score)
Johnson et al., 2018 * [15]	Y	Y	Y	Y	NR	NO	NO	NO	Y	NA	Y	NO	NA	NO	Fair
Trollip et al., 2019 * [16]	Y	Y	Y	Y	NR	NO	NO	NO	Y	NA	Y	NO	NA	NO	Fair
Hurt et al., 2015 * [17]	NO	Y	Y	NR	NO	Y	Y	NO	NO	NO	Y	NO	NA	NO	Poor
Batsis et al., 2020 [18]	Y	Y	Y	Y	NR	Y	Y	Y	Y	Y	Y	NO	Y	NO	Good
Kernizan et al., 2020 [19]	Y	Y	Y	Y	NO	Y	Y	Y	Y	Y	Y	NO	Y	NO	Good
Samela et al., 2017 [20]	Y	Y	Y	Y	NO	Y	Y	Y	Y	Y	Y	NO	Y	NO	Good
Fabiani et al., 2020 [21]	Y	Y	Y	Y	NO	Y	Y	Y	Y	NO	Y	NO	Y	Y	Good

CD: cannot determine; NA: not applicable; NIH: National Institutes of Health; NR: not reported; Y: yes. Q1: Was study question or objective clearly stated? Q2: Was the study population clearly specified and defined? Q3: Was the participation rate of eligible persons at least 50%? Q4: Were all the subjects selected or recruited from the same or similar populations (including the same time period)? Were inclusion and exclusion criteria for being in the study prespecified and applied uniformly to all participants? Q5: Was a sample size justification, power description, or variance and effect estimates provided? Q6: For the analyses in this paper, were the exposure(s) of interest measured prior to the outcome(s) being measured? Q7: Was the timeframe sufficient so that one could reasonably expect to see an association between exposure and outcome if it existed? Q8: For exposures that can vary in amount or level, did the study examine different levels of the exposure as related to the outcome (e.g., categories of exposure, or exposure measured as a continuous variable)? Q9: Were the exposure measures (independent variables) clearly defined, valid, reliable, and implemented consistently across all study participants? Q10: Was the exposure(s) assessed more than once over time? Q11: Were the outcome measures (dependent variables) clearly defined, valid, reliable, and implemented consistently across all study participants? Q12: Were the outcome assessors blinded to the exposure status of participants? Q13: Was loss to follow-up after baseline 20% or less? Q14: Were key potential confounding variables measured and adjusted statistically for their impact on the relationship between exposure(s) and outcome(s)?

3.3. Findings in the Pediatric Population

Among studies analyzing GI symptoms' prevalence in the pediatric population, the Johnson et al. cross-sectional research reported a lower rate of GI symptoms among BTF users (60% of 217) rather than SCF users (97% of 214). The most frequently reported GI problems in children receiving BTF versus SCF was constipation (18.6% vs. 17.8%) followed by vomiting (13.6% vs. 21%), gas/bloating (11.4% vs. 18.3%), diarrhea (5.4% vs. 11.4%), nausea (3.9% vs. 14.8%), pain (3.9% vs. 11.7%), and fever (1.1% vs. 2.15%) [15].

Table 2. Quality Assessment of Included Studies using NIH Quality Assessment Tool for Before-After (Pre-Post) Studies with No Control Group.

Study	Q1	Q2	Q3	Q4	Q5	Q6	Q7	Q8	Q9	Q10	Q11	Q12	Quality (Total Quality Score)
Pentiuk et al., 2011 [22]	Y	Y	Y	Y	NO	Y	Y	NA	Y	NO	Y	NA	Fair
Gallagher et al., 2018 [23]	Y	Y	Y	Y	NO	Y	Y	NO	Y	Y	Y	NA	Good
Hron et al., 2019 [24]	Y	Y	Y	Y	NO	Y	Y	NO	Y	Y	NO	NA	Fair
Spurlock et al., 2022 [25]	Y	Y	Y	Y	NO	Y	Y	NO	NO	Y	Y	NA	Fair

CD: cannot determine; NA: not applicable; NIH: National Institutes of Health; NR: not reported; Y: yes. Q1:Was the study question or objective clearly stated? Q2: Were eligibility/selection criteria for the study population prespecified and clearly described? Q3: Were the participants in the study representative of those who would be eligible for the test/service/intervention in the general or clinical population of interest? Q4: Were all eligible participants that met the prespecified entry criteria enrolled? Q5: Was the sample size sufficiently large to provide confidence in the findings? Q6: Was the test/service/intervention clearly described and delivered consistently across the study population? Q7: Were the outcome measures prespecified, clearly defined, valid, reliable, and assessed consistently across all study participants? Q8: Were the people assessing the outcomes blinded to the participants' exposures/interventions? Q9: Was the loss to follow-up after baseline 20% or less? Were those lost to follow-up accounted for in the analysis? Q10: Did the statistical methods examine changes in outcome measures from before to after the intervention? Were statistical tests done that provided *p* values for the pre-to-post changes? Q11:Were outcome measures of interest taken multiple times before the intervention and multiple times after the intervention (i.e., did they use an interrupted time-series design)? Q12: If the intervention was conducted at a group level (e.g., a whole hospital, a community, etc.) did the statistical analysis take into account the use of individual-level data to determine effects at the group level?

Table 3. Quality Assessment of Relevant Study using NIH Quality Assessment Tool for Controlled Intervention Studies.

Study	Q1	Q2	Q3	Q4	Q5	Q6	Q7	Q8	Q9	Q10	Q11	Q12	Q13	Q14	Quality (Total Quality Score)
Schmidt et al., 2018 [5]	Y	Y	Y	NO	NO	Y	NO	Y	Y	Y	Y	NO	Y	Y	Good

CD: cannot determine; NA: not applicable; NIH: National Institutes of Health; NR: not reported; Y: yes. Q1: Was the study described as randomized, a randomized trial, a randomized clinical trial, or an RCT? Q2: Was the method of randomization adequate (i.e., use of randomly generated assignment)? Q3: Was the treatment allocation concealed (so that assignments could not be predicted)? Q4: Were study participants and providers blinded to treatment group assignment? Q5: Were the people assessing the outcomes blinded to the participants' group assignments? Q6: Were the groups similar at baseline on important characteristics that could affect outcomes (e.g., demographics, risk factors, co-morbid conditions)? Q7: Was the overall drop-out rate from the study at endpoint 20% or lower of the number allocated to treatment? Q8: Was the differential drop-out rate (between treatment groups) at endpoint 15 percentage points or lower? Q9: Was there high adherence to the intervention protocols for each treatment group? Q10: Were other interventions avoided or similar in the groups (e.g., similar background treatments)? Q11: Were outcomes assessed using valid and reliable measures, implemented consistently across all study participants? Q12: Did the authors report that the sample size was sufficiently large to be able to detect a difference in the main outcome between groups with at least 80% power? Q13: Were outcomes reported or subgroups analyzed prespecified (i.e., identified before analyses were conducted)? Q14: Were all randomized participants analyzed in the group to which they were originally assigned, i.e., did they use an intention-to-treat analysis?

In Trollip et al.'s survey, a marked improvement of upper GI symptoms was reported by most caregivers in children after commencing BTF (n = 12). Specifically, the median score obtained through a novel qualitative questionnaire, developed using items from a well-validated scale on feeding assessment in the pediatric age, showed a significant improvement in GI symptoms. Vomiting and nausea frequency changed from 'often' to 'rare', and reflux changed from 'often' to 'rare'. An improvement in aspiration rate was experienced by one-third of the populations analyzed, with only one caregiver reporting a worsening post-BTF initiation. The reported perceived benefit of BTF on clinical outcomes included a positive trend in bowel habits (n = 10), specifically in constipation (from 'often' to 'sometimes') and diarrhea rates (from 'rarely' to 'never'). Abdominal pain rates remained primarily unchanged [16].

Table 4. Features of Included Studies.

Study	Population and Sample Size	Outcome Evaluation Method	Blenderized Food—Type	Diet Prescription Guidelines	Follow-Up Period
Johnson et al., 2018 [15]	255 children 50.5% (n = 173) using CF and 49.5% (n = 82) using BTF	Questionnaire	Homemade (59%) or commercially available blenderized	NR	NA
Trollip et al., 2019 [16]	12 children	Questionnaire	Homemade (33%) or a combination of homemade and formula (17.5%)	NR	NA
Hurt et al., 2015 [17]	54 adults 50.5% (n = 30) using BTF and 45.5% (n = 24) using CF	Self-designed survey	Homemade	NR	NA
Batsis et al., 2020 [18]	23 children	Clinical documentation provided in the medical records	Homemade (65%) or commercially available blenderized (17.5%) or a combination of both (17.5%)	NR	3 and 6 months
Kernizan et al., 2020 [19]	34 children	Parent report Clinical documentation provided in the medical records	Homemade	Recipes designed by dietitians	NA
Samela et al., 2017 [20]	10 children All transitioned from an elemental formula to real food ingredients formula (TFRF)	The number of defecations and the consistency of each stool	Commercially available blenderized	NR	NA
Fabiani et al., 2020 [21]	250 adults 103 fed blenderized natural enteral feeding and 112 fed commercial formulas.	The number of defecations and the consistency of each stool according to the Bristol Stool Chart (BSC)	Homemade	Simple weight-based equation (25–30 kcal/kg/day) to calculate daily caloric target (Atasever et al., 2018; Taylor et al., 2016)	8 days
Pentiuk et al., 2011 [22]	33 children	Survey	Homemade	Food Processor Program (ESHA Research, Salem, OR)	2 months
Gallagher et al., 2018 [23]	20 children	Questionnaire	Homemade	Canada's Food Guide for Healthy Eating. VitamixR 7500 G-Series blender	6 months
Hron et al., 2019 [24]	70 children	Pediatric Gastroesophageal Reflux Disease Symptom and Quality-of-Life Questionnaire (PGSQ); (L) PedsQL Gastrointestinal Symptoms Scale (GI-PedsQL).	Homemade (67%) or commercially available blenderized	Cronometer, a web-based nutrient database (Revelstoke, British Columbia)	1 year
Spurlok et al., 2022 [25]	14 adults first 2 weeks in CF, next 3 weeks partial BTF, next full BTF	Questionnaire	Commercially available blenderized	NR	6 weeks
Schmidt et al., 2019 [5]	118 adults 50% using commercially available product based on real foods and 50% using standard tube feed	The number of defecations and the consistency of each stool according to the Bristol Stool Chart (BSC)	Commercially available blenderized	FAO Expert Consultation on Energy and Protein Requirements (1985)	1 month

NA: not applicable, NR: not reported.

Data on GI symptoms retrospectively collected by Batsis et al. on 23 children switching from SCM to BTF highlighted an overall improvement. The majority of the included population (n = 21/23) while on standard enteral formulas complained about upper GI symptoms, namely gagging (39%), emesis (48%), and chronic cough with concern for aspiration (4%). Ninety-five percent of them experienced an improvement in the reported symptoms over a three-month period. BTFs did not decrease the constipation rate in the patient previously suffering from it, and new onset mild constipation was reported in 21% (n = 5) of patients [18].

Similar data were reported by Kernizan et al. in 35 highly complex patients, almost all (n = 33) suffering from GI symptoms before switching to partial or full BTF. Sixty percent of them (n = 21/35) experienced a gradual improvement of symptoms during the follow-up visits. The most commonly improved symptoms were those related to gastroesophageal reflux disease (GERD). Two patients had worsening of constipation and GERD, respectively [19].

Furthermore, Samela et al. reported a reduction in lower GI symptoms in 10 patients suffering from intestinal failure to absorb macro- and micronutrients, transitioning from SCF to tube feeding formula with real food ingredients (TFRF). Specifically, stooling patterns (consistency or volume and number in 24 h) improved in the majority of cases (90%). TFRF resulted in being well tolerated in children with 30–40 cm of small bowel, an intact ileocecal valve, and at least two-thirds of their colons in continuity [20].

Petniuk et al. reported that fifty-two percent (n = 17/33) of children with fundoplicatio experienced an extensive decrease (76% to 100%) in gagging and retching after two months on BTF. No parents reported that their child's GI symptoms worsened after starting BTF [22].

In a more extended monitoring period of six months for 33 patients, Gallagher et al. recorded a decrease in gagging and retching from 82% pre-BTF to 47% post-BTF. Between enrolment and study exit, stool frequency of more than one/day slightly decreased (from 100% to 94%), while stool consistency did not significantly change [23].

In a larger cohort study of Hron et al. on 70 children, participants receiving blenderized diets compared with those receiving SCF showed fewer GERD symptoms assessed by the Pediatric Gastroesophageal Symptom and the Quality-of-Life Questionnaire (PGSQ). Furthermore, participants on blenderized diets indicated an overall improved gastrointestinal function through Pediatric Quality-of-Life Inventory Gastrointestinal Symptoms (PedsQL). Specifically, less nausea and vomiting (64.0 ± 22.6 vs. 49.0 ± 37.9, p = 0.02), less abdominal pain (65.0 ± 26.8 vs. 56.4 ± 33.9, p = 0.04), abdominal upset (71.1 ± 26.0 vs. 58.9 ± 32.7, p = 0.02), less diarrhea (87.9 ± 15.5 vs. 73.6 ± 26.3, p = 0.004), and less worry about stool (91.5 ± 12.8 vs. 81.4 ± 30.0, p = 0.05) were reported [24].

3.4. Findings in the Adult Population

In Hurt et al.'s survey, among 54 adults using either BTF (n = 30) or CF (n = 24), nausea/vomiting was scarcely reported (13%) in both groups. On the contrary, diarrhea was experienced by 21% of the 24 patients using SCF and by 16% of those using BTF. Constipation was reported by 6% and 3%, respectively, in the BTF and SCF groups [17].

Over an 8-day-observation window, Fabiani et al. found that roughly half of the 112 critically ill patients (due to cardiac surgery) fed with SCF developed diarrhea, while this symptom occurred in less than one-third of the 103 patients fed by BTF [21].

In a cohort of adults with head and neck cancer, Spurlock et al. found that all GI symptoms improved after switching from SCF to BTF. GI symptoms decreased, particularly vomiting (31.3% to 12.5%), constipation (31.3% to 12.5%), gas/bloating (50% to 18.8%), nausea (62.5% to 12.5%), and diarrhea (37.5% to 0%) [25].

Lastly, Schmidt et al. in their RCT provided that in critically ill neurological patients BTF may considerably reduce the number of watery stools and diarrhea, over a 24-day-observation of complete enteral nutrition, when compared to fiber-based SCF [5].

4. Discussion

The sentence "What is old is new again" is particularly true when considering BTF. While is it ancient like old Egyptians, the popularity and necessity of this type of nutrition reduced with the emergence of commercial formulas in the middle 20th century. As technology advanced over the 1960s and 1970s, SCF (for definition, sterile products) began to replace home-prepared food because of their known nutrient composition and lack of possible microbial contamination [8]. However, lately, the interest in these individualized formulas has increased, especially as a request made by caregivers (a way of feeding perceived more "normal" by the family) [10].

Thirty years ago, the estimated annual prevalence of HEN in the USA was 415 per million people [26]. The practice of HEN has faced extensive growth over time, especially in the pediatric population, where the estimated overall prevalence is 3.47 per 100,000 inhabitants from 0 to 18 years of age [27]. This increase is driven by the rising prevalence of feeding and swallowing difficulties connected to improved survival rates of children with complex disabilities and rare genetic conditions [28–32]. In addition to SCF intolerance, one reason of the emerging interest and use of BFT is the inability to obtain commercial formulas in some peculiar settings [33]. To date, in some developing economies, such as Iran, for most hospitals, the traditional blended formulas remain the most widely used option due to higher affordability [34]. Conversely, in the Medicare and Medicaid context, food-based products are a second choice, preferred in case of allergy or intolerance to semi-synthetic formulas [26]. Although SCF guarantees an adequate supply of nutrients, with a low risk of contamination and device obstruction, intolerance has been reported [10]. The improvement of this latter concern is reported in multiple studies both in pediatric and adult populations [8,9,17].

Many aspects have contributed to the re-emergence of BTF. Paramount, for example, is the improvement in feeding tolerance (reduction in reflux, retching/gagging, constipation). This, indeed, greatly improved psychosocial aspects related to feeding, like normalizing mealtimes, allowing patients to participate in food preparation, and allowing caregivers to fulfill the fundamental role of feeding their child [8]. The BLEND study by Pentiuk et al. found a clear improvement in the QoL of families in which BTF was adopted [22].

However, some critical points are to be considered by clinicians before initiating BTF. The patient should be medically stable on a home enteral nutrition regimen, tolerate bolus feedings, and have access to the necessary equipment to prepare and store food. The gastrostomy site should be mature and well-maintained, and the gastrostomy tube should be $\geq$14 French to reduce the risk of tube occlusion [8]. Nonetheless, there are some concerns about BTF. Food-borne illness, related to the preparation of a "whole food formula", is a possible "side effect" that commonly leads clinicians to prefer commercial formula and represents a reason of concern for patients/caregivers, particularly for critically ill or immunocompromised patients, such as neonates [35].

Regardless, no available studies demonstrate a connection between higher levels of bacterial contamination and increased infection rates in BTF [36]. Additionally, a 2020 study by Milton et al. shows that, when a correct way of preparation of BTF is applied, 88% of samples meet the criteria for safe food consumption [37].

With regard to contamination concerns, many authors suggest the use of accepted methods of safe food handling [37], adequate hygiene measures, and the use of comprehensive guidelines for preparation, storage, and transportation of BTF [38,39].

A review based on the adult population showed that BTFs are inappropriate for use in medically complex patients or those at risk for malnutrition, since BTF seems to be associated with lower nutrient adequacy, possibly leading to a decline in weight status, BMI, and upper arm circumference [40]. This aspect could be of concern in more fragile pediatric patients.

Overall volume tolerance should also be factored in. Given the patient's level of volume sensitivity, the patient could not meet the daily caloric and nutrient intake needed with BTF alone, which is crucial in children's growth [8]. The BLEND3 study found that children needed 1.5-fold calories when on BTF in comparison to SCF to sustain growth. It is still controversial why a caloric increase is necessary for BTF. Possible explanations include differences in thermic effects of feeding, miscalculation of the caloric value of foods, or changes in food digestion on BTF. However, further investigation is needed to clarify this point [36].

The most relevant evidence emerging from this research study is that all included papers unanimously converge in defining an improvement in both upper and lower GI symptoms during BTF both in the pediatric and adult populations regardless of their medical conditions. GI symptoms, including diarrhea and abdominal distension, frequently occur in patients receiving EN, and diverse causes, such as antibiotics, infections, or even enteral nutrition, may contribute [41].

All included studies showed comparable results, indicating an improvement of HEN tolerance between patients using BTF. Notwithstanding, experimental studies, particularly RCTs, are lacking. Adequate methodological quality is only sometimes achieved in the included studies in this review. In fact, dietetic prescription, as well as concomitant possible antibiotic administration, are not always described owing to observational biases. To note, blinding is more difficult to achieve in studies on feeds as less feasible. Additionally, due to small cohorts, statistical power is lacking.

Despite all the limitations, this review highlights the possible improvement of BTF on clinical outcomes. It fits into a stream of studies covering most aspects of HEN, including future strategies to improve environmental sustainability of HEN [42]. Results of studies on nutritional value, quality of life, and microbial contamination of BTF are not always consistent, as they do not always show that BTF is more beneficial than SCF.

Notably, BTF is not a good alternative in patients requiring jejunostomy in order to prevent metabolic complications. Since this feeding route requires the administration of feeds through a feeding pump, BTF is not recommended [1,43–45].

To conclude, there is a strong consensus agreement on most the appropriate HEN formula among experts. ESPEN guideline recommends the use of SCF as the first choice, unless there is the presence of a specific justification for BTF [1]. The absence of standardized BTF formulas may potentially increase the risk of malnutrition due to deficiencies of micronutrients [9,44] that are fundamental for effective metabolism and biochemical processes [46,47]. Hence, the importance of an appropriate interdisciplinary management of BTF, especially by an expert dietitian. A 3-day food diary may be filled in to estimate the adequate supply of macro- and micronutrients and to prevent weight loss.

5. Conclusions

Empirically, BTF is sometimes preferred to the more conventional CTF because of the emergence of GI symptoms during enteral nutrition with CTF. Although several studies conducted on adults and pediatrics report an improvement in GI symptoms' frequency during BTF in comparison with CTF, only a few studies report with a high degree of methodological quality. Considering these findings, experimental research is needed to develop most robust evidence on this topic that is gaining increasing consideration among caregivers and patients.

6. Future Directions

Despite the difficulties with blinding and conducting RCTs in feeding studies, more studies are needed to evaluate the effectiveness of blenderized tube feedings with prolonged follow-up period and adequate medical monitoring to ensure optimal delivery and nutritional standards. In the future, it will be interesting to consider the economic and eco-sustainable impact of BTF in addition to the evaluation of clinical competence.

Supplementary Materials: The following supporting information can be downloaded at: https://www.mdpi.com/article/10.3390/app13042173/s1, Table S1: Methodology of search for articles evaluated in this Review; Table S2: Preferred Reporting Items for Systematic reviews and Meta-Analyses extension for Scoping Reviews (PRISMA-ScR) Checklist, From [12].

Author Contributions: Study concept and design: R.O., E.S., D.L. and G.Z. Acquisition of data: E.S. and R.O. Analysis and interpretation of data: E.S. and D.L. Drafting the manuscript: E.S. and R.O. Critical revision of the manuscript for important intellectual content: D.R., V.G., E.R. and G.Z. Conceptualization and editing of the manuscript: R.O., E.S., D.L., V.G., G.M., F.P., E.M.K., C.L., D.R., I.C., C.V., E.R., A.G. and G.Z. All authors have read and agreed to the published version of the manuscript.

Funding: This research received no external funding.

Institutional Review Board Statement: Not applicable.

Informed Consent Statement: Not applicable.

Data Availability Statement: Not applicable.

Conflicts of Interest: The authors declare no competing interest.

Abbreviations

BTF	blenderized tube feeding
ESPEN	European Society for Clinical Nutrition and Metabolism
ETF	enteral tube feeding
G-tube	gastrostomy tube
HEN	home enteral nutrition
SCF	standard commercial formula

References

1. Bischoff, S.C.; Austin, P.; Boeykens, K.; Chourdakis, M.; Cuerda, C.; Jonkers-Schuitema, C.; Lichota, M.; Nyulasi, I.; Schneider, S.M.; Stanga, Z.; et al. ESPEN guideline on home enteral nutrition. *Clin. Nutr.* **2020**, *39*, 5–22. [CrossRef]
2. Gramlich, L.; Hurt, R.T.; Jin, J.; Mundi, M.S. Home Enteral Nutrition: Towards a Standard of Care. *Nutrients* **2018**, *10*, 1020. [CrossRef]
3. Bering, J.; DiBaise, J.K. Home Parenteral and Enteral Nutrition. *Nutrients* **2022**, *14*, 2558. [CrossRef]
4. Doley, J. Enteral Nutrition Overview. *Nutrients* **2022**, *14*, 2180. [CrossRef]
5. Schmidt, S.B.; Kulig, W.; Winter, R.; Vasold, A.S.; Knoll, A.E.; Rollnik, J.D. The effect of a natural food based tube feeding in minimizing diarrhea in critically ill neurological patients. *Clin. Nutr.* **2019**, *38*, 332–340. [CrossRef]
6. Oparaji, J.A.; Sferra, T.; Sankararaman, S. Basics of Blenderized Tube Feeds: A Primer for Pediatric Primary Care Clinicians. *Gastroenterol. Res.* **2019**, *12*, 111–114. [CrossRef]
7. O'Flaherty, T.; Santoro, K.; Pentiuk, S. Calculating and Preparing a Pureed-by-Gastrostomy-Tube (PBGT) Diet for Pediatric Patients with Retching and Gagging Postfundoplication. *Infant Child Adolesc. Nutr.* **2011**, *3*, 361–364. [CrossRef]
8. Bobo, E. Reemergence of Blenderized Tube Feedings: Exploring the Evidence. *Nutr. Clin. Pract.* **2016**, *31*, 730–735. [CrossRef]
9. Epp, L.; Lammert, L.; Vallumsetla, N.; Hurt, R.T.; Mundi, M.S. Use of Blenderized Tube Feeding in Adult and Pediatric Home Enteral Nutrition Patients. *Nutr. Clin. Pract.* **2017**, *32*, 201–205. [CrossRef]
10. Bennett, K.; Hjelmgren, B.; Piazza, J. Blenderized Tube Feeding: Health Outcomes and Review of Homemade and Commercially Prepared Products. *Nutr. Clin. Pract.* **2020**, *35*, 417–431. [CrossRef]
11. Ribeiro-Mourão, F.; Bertaud, S.; Brierley, J.; McCulloch, R.; Köglmeier, J.; Hill, S.M. Use of home parenteral nutrition in severely neurologically impaired children. *Arch. Dis. Child* **2022**, *107*, 365–370. [CrossRef]
12. Tricco, A.C.; Lillie, E.; Zarin, W.; O'Brien, K.K.; Colquhoun, H.; Levac, D.; Moher, D.; Peters, M.D.J.; Horsley, T.; Weeks, L.; et al. PRISMA Extension for Scoping Reviews (PRISMA-ScR): Checklist and Explanation. *Ann. Intern. Med.* **2018**, *169*, 467–473. [CrossRef]
13. NHLBI. Available online: https://www.nhlbi.nih.gov/health-topics/study-quality-assessment-tools (accessed on 1 September 2022).
14. Ma, L.-L.; Wang, Y.-Y.; Yang, Z.-H.; Huang, D.; Weng, H.; Zeng, X.-T. Methodological quality (risk of bias) assessment tools for primary and secondary medical studies: What are they and which is better? *Mil. Med. Res.* **2020**, *7*, 7. [CrossRef]
15. Johnson, T.W.; Milton, D.L.; Johnson, K.; Carter, H.; Hurt, R.T.; Mundi, M.S.; Epp, L.; Spurlock, A.L. Comparison of Microbial Growth Between Commercial Formula and Blenderized Food for Tube Feeding. *Nutr. Clin. Pract.* **2019**, *34*, 257–263. [CrossRef]
16. Trollip, A.; Lindeback, R.; Banerjee, K. Parental Perspectives on Blenderized Tube Feeds for Children Requiring Supplemental Nutrition. *Nutr. Clin. Pract.* **2020**, *35*, 471–478. [CrossRef]
17. Hurt, R.T.; Varayil, J.E.; Epp, L.M.; Pattinson, A.K.; Lammert, L.M.; Lintz, J.E.; Mundi, M.S. Blenderized Tube Feeding Use in Adult Home Enteral Nutrition Patients: A Cross-Sectional Study. *Nutr. Clin. Pract.* **2015**, *30*, 824–829. [CrossRef]
18. Batsis, I.D.; Davis, L.; Prichett, L.; Wu, L.; Shores, D.; Yueng, K.A.; Oliva-Hemker, M. Efficacy and Tolerance of Blended Diets in Children Receiving Gastrostomy Feeds. *Nutr. Clin. Pract.* **2020**, *35*, 282–288. [CrossRef]

19. Kernizan, D.; Mintz, D.; Colin, M.; Melanie, L.; Lindsay, Y.; Yen Ping, C.; Elizaveta, I.; Soula, K.; Melissa, W. Outcomes and Safety of Blenderized Tube Feedings in Pediatric Patients: A Single Center's Experience. *J. Pediatr. Gastroenterol. Nutr.* **2020**, *71*, e124–e128. [CrossRef]
20. Samela, K.; Mokha, J.; Emerick, K.; Davidovics, Z.H. Transition to a Tube Feeding Formula with Real Food Ingredients in Pediatric Patients With Intestinal Failure. *Nutr. Clin. Pract.* **2017**, *32*, 277–281. [CrossRef]
21. Fabiani, A.; Sanson, G.; Bottigliengo, D.; Dreas, L.; Zanetti, M.; Lorenzoni, G.; Gatti, G.; Sacilotto, M.; Pappalardo, A.; Gregori, D. Impact of a natural versus commercial enteral-feeding on the occurrence of diarrhea in critically ill cardiac surgery patients. A retrospective cohort study. *Int. J. Nurs. Stud.* **2020**, *108*, 103605. [CrossRef]
22. Pentiuk, S.; O'Flaherty, T.; Santoro, K.; Willging, P.; Kaul, A. Pureed by gastrostomy tube diet improves gagging and retching in children with fundoplication. *J. Parenter. Enteral. Nutr.* **2011**, *35*, 375–379. [CrossRef]
23. Gallagher, K.; Flint, A.; Mouzaki, M.; Carpenter, A.; Haliburton, B.; Bannister, L.; Norgrove, H.; Hoffman, L.; Mack, D.; Stintzi, A.; et al. Blenderized Enteral Nutrition Diet Study: Feasibility, Clinical, and Microbiome Outcomes of Providing Blenderized Feeds Through a Gastric Tube in a Medically Complex Pediatric Population. *J. Parenter. Enteral. Nutr.* **2018**, *42*, 1046–1060. [CrossRef]
24. Hron, B.; Fishman, E.; Lurie, M.; Clarke, T.; Chin, Z.; Hester, L.; Burch, E.; Rosen, R. Health Outcomes and Quality of Life Indices of Children Receiving Blenderized Feeds via Enteral Tube. *J. Pediatr.* **2019**, *211*, 139–145. [CrossRef] [PubMed]
25. Spurlock, A.Y.; Johnson, T.W.; Pritchett, A.; Pierce, L.; Hussey, J.; Johnson, K.; Carter, H.; Davidson, S.L.; Mundi, M.S.; Epp, L.; et al. Blenderized food tube feeding in patients with head and neck cancer. *Nutr. Clin. Pract.* **2022**, *37*, 615–624. [CrossRef]
26. Howard, L.; Ament, M.; Fleming, C.R.; Shike, M.; Steiger, E. Current use and clinical outcome of home parenteral and enteral nutrition therapies in the United States. *Gastroenterology* **1995**, *109*, 355–365.
27. Diamanti, A.; Di Ciommo, V.M.; Tentolini, A.; Lezo, A.; Spagnuolo, M.I.; Campanozzi, A.; Panetta, F.; Basso, M.S.; Elia, D.; Gambarara, M. Home enteral nutrition in children: A 14-year multicenter survey. *Eur. J. Clin. Nutr.* **2013**, *67*, 53–57. [CrossRef]
28. American Speech-Language-Hearing Association. Pediatric Feeding and Swallowing. (Practice Portal). Available online: https://www.asha.org/practice-portal/clinical-topics/pediatric-feeding-and-swallowing/ (accessed on 1 October 2022).
29. Onesimo, R.; Giorgio, V.; Viscogliosi, G.; Sforza, E.; Kuczynska, E.; Margiotta, G.; Iademarco, M.; Proli, F.; Rigante, D.; Zampino, G.; et al. Management of nutritional and gastrointestinal issues in RASopathies: A narrative review. *Am. J. Med. Genet. C Semin. Med. Genet.* **2022**, *190*, 478–493. [CrossRef]
30. Giorgio, V.; Margiotta, G.; Stella, G.; Cicco, F.D.; Leoni, C.; Proli, F.; Zampino, G.; Gasbarrini, A.; Onesimo, R. Intestinal Permeability in Children with Functional Gastrointestinal Disorders: The Effects of Diet. *Nutrients* **2022**, *14*, 1578. [CrossRef]
31. Parma, B.; Cianci, P.; Mariani, M.; Cereda, A.; Panceri, R.; Fossati, C.; Maestri, L.; Macchini, F.; Onesimo, R.; Zampino, G.; et al. Nissen fundoplication in Cornelia de Lange syndrome spectrum: Who are the potential candidates? *Am. J. Med. Genet. A* **2020**, *182*, 1697–1703. [CrossRef]
32. Sforza, E.; Onesimo, R.; Leoni, C.; Giorgio, V.; Proli, F.; Notaro, F.; Kuczynska, E.M.; Cerchiari, A.; Selicorni, A.; Rigante, D.; et al. Drooling outcome measures in paediatric disability: A systematic review. *Eur. J. Pediatr.* **2022**, *181*, 2575–2592. [CrossRef]
33. Johnson, T.W.; Spurlock, A.; Pierce, L. Survey study assessing attitudes and experiences of pediatric registered dietitians regarding blended food by gastrostomy tube feeding. *Nutr. Clin. Pract.* **2015**, *30*, 402–405. [CrossRef] [PubMed]
34. Jazayeri, S.M.H.; Ostadrahimi, A.; Safaiyan, A.; Hashemzadeh, S.; Salehpour, F. Standard Enteral Feeding Improves Nutritional Status Compared with Hospital-Prepared Blended Formula Among Intensive Care Unit (ICU) Patients. *Prog. Nutr.* **2016**, *18*, 22–25. Available online: https://mattioli1885journals.com/index.php/progressinnutrition/article/view/4308 (accessed on 1 October 2022).
35. Steel, C.; Wile, H.; Li, O.; Yedulla, S.; Hare, I.; Hopkins, B. Understanding the use and tolerance of a pediatric and an adult commercial blenderized enteral formula through real-world data. *Nutr. Clin. Pract.* **2022**, 1–9. [CrossRef]
36. Chandrasekar, N.; Dehlsen, K.; Leach, S.T.; Krishnan, U. Exploring Clinical Outcomes and Feasibility of Blended Tube Feeds in Children. *J. Parenter. Enteral. Nutr.* **2021**, *45*, 685–698. [CrossRef]
37. Milton, D.L.; Johnson, T.W.; Johnson, K.; Murphy, B.; Carter, H.; Hurt, R.T.; Mundi, M.S.; Epp, L.; Spurlock, A.Y.; Hussey, J. Accepted Safe Food-Handling Procedures Minimizes Microbial Contamination of Home-Prepared Blenderized Tube-Feeding. *Nutr. Clin. Pract.* **2020**, *35*, 479–486. [CrossRef]
38. Galindo, C.O.; Beux, M.R.; da Costa, R.L.; Uniat, K.C.; Leobet, J.; Ferreira, S.M.R.; Medeiros, C.O.; Schieferdecker, M.E.M.; Stangarlin-Fiori, L. Home-Prepared Enteral Tube Feeding: Evaluation of Microbiological Contamination, Hygiene, and the Profile of the Food Handler. *Nutr. Clin. Pract.* **2021**, *36*, 704–717. [CrossRef]
39. Ojo, O.; Adegboye, A.R.A.; Ojo, O.O.; Wang, X.; Brooke, J. The Microbial Quality and Safety of Blenderised Enteral Nutrition Formula: A Systematic Review. *Int. J. Environ. Res. Public Health* **2020**, *17*, 9563. [CrossRef]
40. Brown, T.; Zelig, R.; Radler, D.R. Clinical Outcomes Associated with Commercial and Homemade Blenderized Tube Feedings: A Literature Review. *Nutr. Clin. Pract.* **2020**, *35*, 442–453. [CrossRef]
41. Barrett, J.S.; Shepherd, S.J.; Gibson, P.R. Strategies to manage gastrointestinal symptoms complicating enteral feeding. *J. Parenter. Enteral. Nutr.* **2009**, *33*, 21–26. [CrossRef]
42. Wayne, N.; Solomon, J.; Martinez Moreno, F.J.; Kocinski, M.; Wyk, J.V.; Hubbard, G.P. A review of home enteral tube feeding home care service and product provision, with the aim to improve environmental sustainability. *Clin. Nutr. ESPEN* **2020**, *35*, 208–252.

43. Zettle, S. Deconstructing Pediatric Blenderized Tube Feeding: Getting Started and Problem Solving Common Concerns. *Nutr. Clin. Pract.* **2016**, *31*, 773–779. [CrossRef] [PubMed]
44. Bischoff, S.C.; Austin, P.; Boeykens, K.; Chuordakis, M.; Cuerda, C.; Jonkers-Schuitema, C.; Lichota, M.; Nyulasi, I.; Schneider, S.M.; Stanga, Z.; et al. ESPEN practical guideline: Home enteral nutrition. *Clin. Nutr.* **2022**, *41*, 468–488. [CrossRef] [PubMed]
45. Boullata, J.I.; Carrera, A.L.; Harvey, L.; Escuro, A.A.; Hudson, L.; Mays, A.; McGinnis, C.; Wessel, J.J.; Bajpai, S.; Beebe, M.L.; et al. ASPEN Safe Practices for Enteral Nutrition Therapy [Formula: See text]. *J. Parenter. Enteral. Nutr.* **2016**, *41*, 15–103. [CrossRef]
46. Kozeniecki, M.; Fritzshall, R. Enteral Nutrition for Adults in the Hospital Setting. *Nutr. Clin. Pract.* **2015**, *30*, 634–651. [CrossRef] [PubMed]
47. Giles, K.H.; Kubrak, C.; Baracos, V.E.; Olson, K.; Mazurak, V.C. Recommended European Society of Parenteral and Enteral Nutrition protein and energy intakes and weight loss in patients with head and neck cancer. *Head Neck* **2016**, *38*, 1248–1257. [CrossRef]

applied sciences

MDPI

Brief Report

Otolaryngology Conditions and Diseases in Migrants: The Experience of the PROTECT Project

Massimo Ralli [1,*], Andrea Colizza [1], Francesca Yoshie Russo [1], Gaspare Palaia [2], Diletta Angeletti [1], Alice Bruscolini [1], Alessia Marinelli [1], Maurizio Bossù [2], Livia Ottolenghi [2], Marco de Vincentiis [1], Antonio Greco [1] and Antonella Polimeni [2]

[1] Department of Sense Organs, Sapienza University of Rome, 00185 Rome, Italy
[2] Department of Oral and Maxillofacial Sciences, Sapienza University of Rome, 00161 Rome, Italy
* Correspondence: massimo.ralli@uniroma1.it

Abstract: Introduction: The number of migrants living in Europe is growing rapidly. The PROTECT project is a national study conducted among refugees and migrants in the Lazio region, Italy from February 2018 to September 2021 to evaluate their otolaryngological, dental, and ophthalmic health status. This article reports the results of this study with a special focus on otolaryngological conditions and diseases. Materials and methods: A total of 3023 participants were included in the project. Data on the participants' demographic factors, migration status, and otolaryngological history were collected using a questionnaire. Each patient underwent clinical ear, nose, and throat examination, and the main otolaryngological conditions and diseases were noted. Results: nearly two-thirds of participants were males (68.1%). The mean age was 31.6 ± 13.1. Most of the participants were born in Nigeria, followed by Bangladesh, Pakistan, Somalia, Mali, and Gambia. The prevalence of chronic noise exposure was 5.2%. Unilateral hearing loss was reported by 6.5% of the subjects, and bilateral hearing loss by 3.6%. The most frequent symptoms reported in the questionnaire were snoring (10.4%), nasal obstruction (5.9%), vertigo (5.0%), otalgia (4.5%), and tinnitus (4.2%). At the clinical examination, the most frequent findings were nasal septum deviation (25.2%), ear wax (6.5%), hypertrophic palatine tonsils (5.3%), and tympanic membrane perforation (1.3%). Conclusions: the PROTECT project allowed for the evaluation of otolaryngological, dental, and ophthalmological conditions in over 3000 migrants, giving them the possibility to access specialist care.

Keywords: migrants; vulnerability; hospitality; otolaryngology; screening; inclusion

Citation: Ralli, M.; Colizza, A.; Russo, F.Y.; Palaia, G.; Angeletti, D.; Bruscolini, A.; Marinelli, A.; Bossù, M.; Ottolenghi, L.; de Vincentiis, M.; et al. Otolaryngology Conditions and Diseases in Migrants: The Experience of the PROTECT Project. *Appl. Sci.* **2023**, *13*, 2104. https://doi.org/10.3390/app13042104

Academic Editor: Bruno Chrcanovic

Received: 13 December 2022
Revised: 31 January 2023
Accepted: 3 February 2023
Published: 6 February 2023

1. Introduction

Persons with migrant status living in Europe are increasing rapidly [1]. European countries receive an increasing number of vulnerable people every year, such as minors, unaccompanied foreign minors, disabled people, pregnant women, single parents with minor children, victims of trafficking, people with illnesses or mental disorders, and people who have been tortured, raped, or suffered other forms of psychological, physical, sexual violence [1,2].

Different factors lead to define migrants as vulnerable subjects, including health risks during and after migration, language, socioeconomical barriers, and difficulty in accessing health services in host countries [3]. Many migrants come from war zones and have been victims of persecutory acts in their country of origin [4].

The rates of migration to Europe have increased in recent years with considerable implications for national health systems [5]. The high costs for treatment in private facilities and barriers to accessing free health services are often the main causes of a high incidence of several diseases in migrant patients, including ear, nose, and throat conditions [6–12]. In fact, the quality of life of these persons is severely impacted by hearing difficulties, breathing issues, and poor oral health, as they could interfere with daily activities such as hearing, eating, and speaking. Furthermore, untreated aural, nasal, or pharyngeal problems could lead to serious risks to the health of older people such as malnutrition, heart disease, pneumonia, oral and pharyngeal cancers, and dementia [13–15]. The recent coronavirus disease 19 (COVID-19) pandemic has further worsened access to healthcare especially for vulnerable persons [16–20].

The "Patologie del distRettO TEsta-Collo nei migranti" (PROTECT) project was started in 2018 and led by the Sapienza University of Rome with the aim of protecting the health of vulnerable migrants, with a specific focus on otolaryngological, dental, and ophthalmic conditions through visits performed by otolaryngologists, dentists, and ophthalmologists. The PROTECT project was funded by the Asylum, Migration, and Integration Fund (FAMI) 2014–2020 of the Italian Ministry of the Interior, and cofinanced by the European Union.

In this article, we describe the results obtained in the PROTECT project with a specific focus on otolaryngological conditions and diseases. The results of this project for other conditions were reported elsewhere [21].

2. Materials and Methods

This is a national cross-sectional study (PROTECT project) conducted on the refugee and migrant population present in the Lazio region, Italy, from February 2018 to September 2021. The otolaryngological health of 3023 participants was investigated in a network of 53 reception centres and integration services. Participants were recruited on site through local refugee and migrant services.

The PROTECT project was approved by the Department of Oral and Maxillofacial Sciences, Sapienza, University of Rome (protocol identifying number: 0000839, 02/10/2018). The protocol was in accordance with the 1975 Declaration of Helsinki on medical protocols and ethics, and its later amendments.

2.1. Data Collection

Data were collected through an interviewer-administered questionnaire and a full clinical otolaryngological examination for each patient. The questionnaire investigated demographic factors, migration status, history for otolaryngological diseases, and health-related behaviours. The clinical examination was performed with the help of an otoscope, a nasal speculum, a laryngeal mirror, and a headlight.

Questionnaire administration and otolaryngological examinations were performed by trained otolaryngological specialists of the Department of Sense Organs, Sapienza University of Rome. The visits took place both at the premises of the reception centres distributed in the Lazio region with the use of mobile medical units and in the outpatient services of the Policlinico Umberto I hospital, Rome, Italy.

2.2. Questionnaire Information

The questionnaire was administered individually through an interview. The questionnaire questions are reported in Table 1. The physician filled the form on the basis of the responses received by each patient during the interview.

Table 1. Individually administered questionnaire to each participant through an operator-administered interview.

(A) Sociodemographic characteristics	
- Full name	
- Gender	
- Date of birth	
- Country of origin	
(B) Otolaryngological anamnesis and health-related behaviours	
- Do you have cases of severe hearing loss in your family?	Yes/no
- Were you exposed to noise for long periods?	Yes/no
- Do you have unilateral hearing loss?	Yes/no
- Do you have bilateral hearing loss?	Yes/no
- Have you ever had injury or trauma in the region of the ear?	Yes/no
- Have you ever had injury or trauma in the region of the nose?	Yes/no
- Have you ever had injury or trauma in the region of the larynx?	Yes/no
- Do you have ear pain?	Yes/no
- Do you have nasal obstruction?	Yes/no
- Do you snore at night?	Yes/no
- Has your voice changed recently?	Yes/no
- Do you have vertigo?	Yes/no
- Do you have tinnitus?	Yes/no

2.3. Clinical Examination

The subjects were examined in the premises of the reception centres distributed in the Lazio region and in the outpatient services of the Policlinico Umberto I hospital of Rome, Italy. For each patient, the otolaryngological specialist evaluated the presence of external ear diseases (malformation of the auricle, malformation of the external ear canal, ear wax, external otitis), tympanic membrane perforation, middle otitis, nasal septum deviation or perforation, acute tonsillitis or peritonsillar abscess, and the presence of spontaneous nystagmus. A standard form listing the above conditions was available for each patient, which was filled by the otolaryngological specialist after the visit using a yes/no response.

In the case of patients who had reported hearing loss at questionnaire, a hearing exam with pure tone audiometry and, if indicated, otoacoustic emissions was performed in the Policlinico Umberto I hospital; the response of the patient was considered valid only if the exams confirmed the presence of hearing loss [22–24]. Hearing loss was defined as a hearing threshold for two or more frequencies >25 dB HL in one or both ears.

If necessary, on the basis of the health evaluation and clinical history, patients were referred to the otolaryngological department of the Policlinico Umberto I hospital for further diagnostic exams or treatments.

2.4. Statistical Analysis

The forms (questionnaire and clinical examination) filled by the examiner during each visit were collected and entered into a database created using Microsoft Excel (Microsoft, Redmond, WA, USA).

Descriptive statistics, including mean ± SD values and percentages, were calculated for each variable, and used to define the main clinical and demographic characteristics.

Data were evaluated using statistical analysis software (version 20.0, Statistical Package for the Social Sciences, IBM Corporation, Armonk, NY, USA).

3. Results

A total of 3023 participants were examined and interviewed. Their age varied from 0 to 93 years; the mean age was 31.6 ± 13.1, SD = 13,075. Among all the subjects, 2058 were male (68.1%) and 965 were female (31.9%). Data are shown in Table 2.

Table 2. Demographic characteristics of migrants included in the study.

		Number	Percentage
Migrants	Male	2058	68.1%
	Female	965	31.9%
	Total	3023	100%

Most of the participants were born in Nigeria (10.15%, n = 307), followed by Bangladesh (7.51%, n = 227), Pakistan (5.45%, n = 165), Somalia (4.76%, n = 144), Mali (4.03%, n = 122), and Gambia (3.97%, n = 120). The average stay in local refugee and migrant services for patients included in the study was 36.2 days (range 1–91, SD = 21,320).

Figure 1 shows the 20 most common countries of origin of the participants.

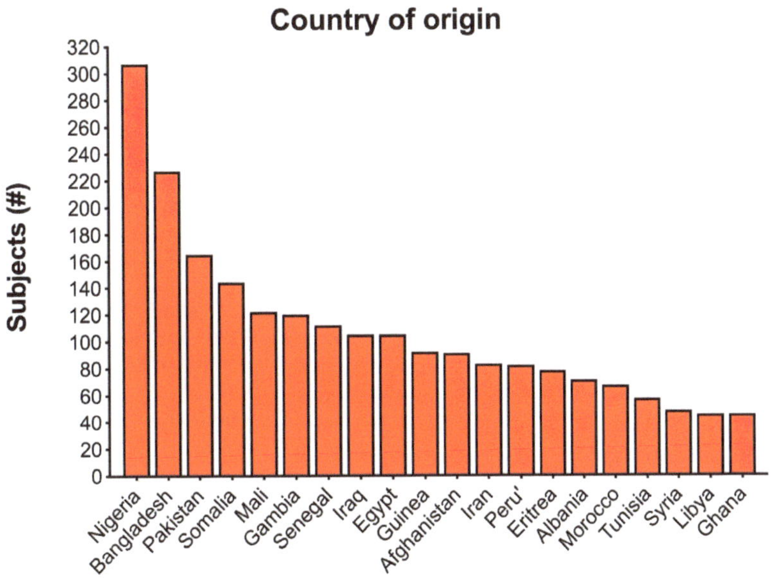

Figure 1. Distribution of the country of origin of the migrants screened in this project sorted by number of patients. Only the first 20 countries are shown.

The most common otolaryngological issue reported in the questionnaire was snoring, present in 10.38% (n = 314) of the subjects. Other conditions reported by the patients were audiologically confirmed unilateral hearing loss (6.48%, n = 196), nasal obstruction (5.89%, n = 178), chronic noise exposure (5.19%, n = 157), subjective vertigo (4.99%, n = 151), nasal trauma (4.89%, n = 148), familial deafness (4.80%, n = 145), ear pain (4.50%, n = 136), tinnitus (4.17%, n = 126), audiologically confirmed bilateral hearing loss (3.60%, n = 109), ear trauma (3.57%, n = 108), laryngeal trauma (0.89%, n = 27), and dysphonia (0.79%, n = 24). Figure 2 details the otolaryngological conditions and symptoms reported by patients in the questionnaire.

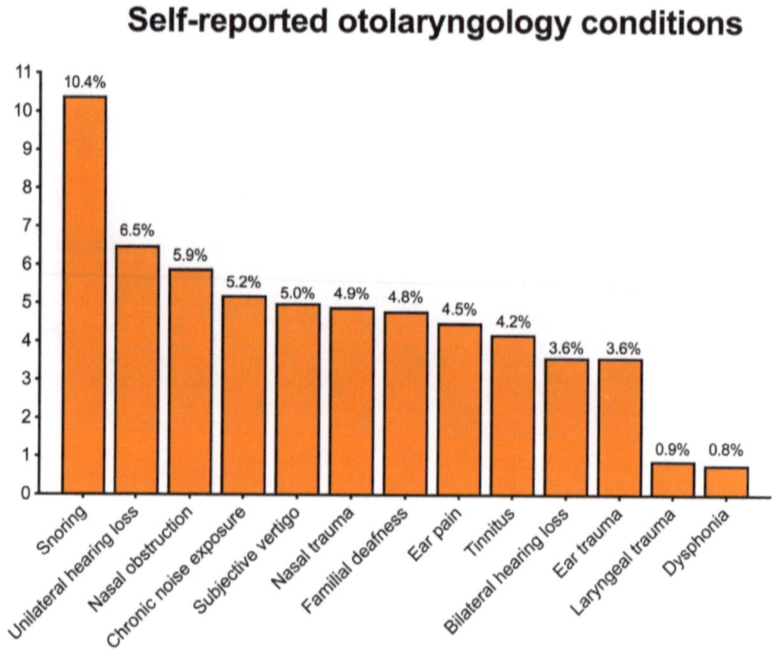

Figure 2. Histogram of otolaryngological conditions and symptoms reported by patients in the operator-administered questionnaire. Data above each column were rounded to the first decimal.

Otolaryngological examination showed a deviation of the nasal septum in 25.21% (n = 762), ear wax in 6.48% (n = 196), and tonsillar hypertrophy in 5.29% of the patients (n = 160). Other less common pathological findings were tympanic membrane perforation (1.29%, n = 39), external otitis (0.89%, n = 27), and middle ear otitis (0.50%, n = 15). Figure 3 shows the otolaryngological conditions found during the clinical examination.

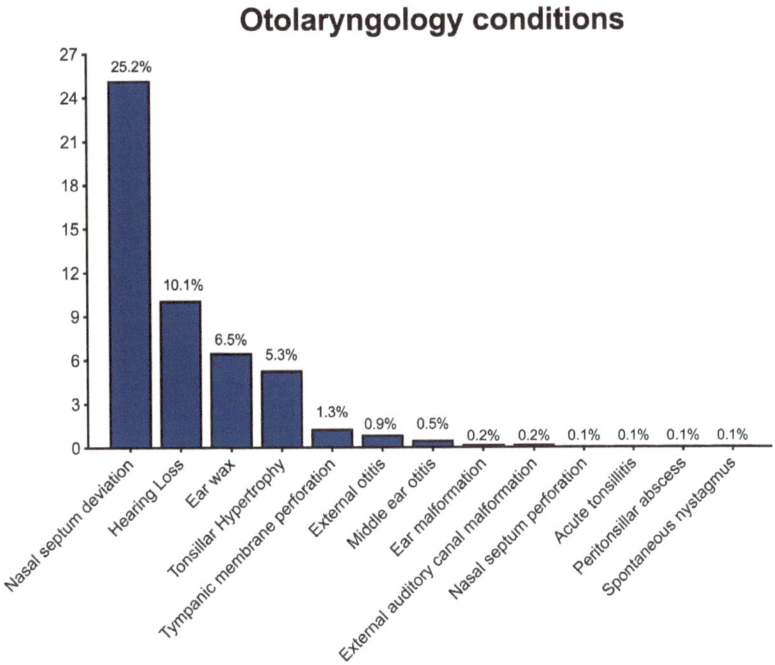

Figure 3. Histogram of otolaryngological conditions found during the clinical examination. Data above each column were rounded to the first decimal.

4. Discussion

In recent years, the prevalence of aural, nasal, or laryngeal diseases in industrialised countries has decreased due to the large use of preventive measures [25,26]. However, prevention is less diffused in nonindustrialised countries, resulting in a higher incidence of common otolaryngological diseases, as found in the sample of migrants included in this study [6,7,10].

Among aural disorders, hearing loss was the most common condition reported during the screening of the migrants included in the study. In the world, one-third of the population over 65 suffer from disabling hearing impairments [27–29]. In Italy, the incidence of hearing loss among the population in the third decade of life, like the mean age of subjects evaluated in this study, is about of 3.2% according to the Italian Society of Occupational Medicine [30]. This number is significantly lower compared to the self-reported hearing loss of patients included in the present study, which involved 6.5% of subjects for unilateral hearing loss and 3.6% for bilateral hearing loss. Differently from the countries of origin of our patients, many factors contributed to the reduction in hearing loss over the last decade in industrialised countries. For example, one of the factors for the pathogenesis of otosclerosis are viral infections [31], and the measles virus shows an important organotropism for the otic capsule [32]. From the 1970s, the introduction of measles vaccination reduced the incidence of otosclerosis [33] and generally of hearing loss [34].

Occupational and environmental noise exposure accounts for the 16% of total hearing loss cases [35–39]. Global governmental bodies that have established regulations to avoid noise exposure include the Occupational Safety and Health Administration, the National Institutes of Occupational Safety and Health, and the European Union (EU) [40]. Unfortunately, these recommendations often are not implemented by the migrants' countries of origin, as shown by the elevated number of persons in our sample reporting hearing loss or chronic noise exposure in their country of origin.

Research conducted in France and the USA showed that, during screening visits among migrants and homeless individuals, the most frequent symptoms involved the upper and lower respiratory tract, such as a dry or productive cough, rhinorrhea, and dyspnea. Sometimes these symptoms may precede life-threatening diseases such as pulmonary infection or tuberculosis [41–43]. Therefore, the possibility of guaranteeing free access to screening the upper airway to all patients is fundamental to identify alterations in nasal airflow (nasal septum deviation or perforation) or chronic infections of the nose, nasopharynx, or oropharynx that could lead to severe infections of the lower airway in the medium and long term.

Quality of life is strongly influenced by the health of the ear, nose and throat. For this reason, it is necessary to render otolaryngological services easily accessible to vulnerable persons and migrants [44]. Projects such as PROTECT aim at improving the quality of life of these patients through the prevention and screening of common conditions, leading to the resolution of the symptoms of an inflammatory/infectious nature or the early interception of more serious conditions such as head and neck neoplasms.

5. Conclusions

The PROTECT project has allowed for the screening of otolaryngological, dental, and ophthalmological conditions in over 3000 migrants, giving them the possibility to access specialist care. Furthermore, for operators of the centers, nurses, and physicians, this project represents an unrepeatable opportunity to offer care to vulnerable persons. The knowledge, prevention, and management of these diseases, which are often painful and disabling in these patients, have led to an improvement in the quality of life of vulnerable migrants included in the PROTECT project.

Author Contributions: Conceptualisation, M.R.; methodology, A.C. and M.R.; validation, A.G. and M.B.; formal analysis, F.Y.R.; investigation, A.M., D.A. and A.M.; resources, A.P. and M.B.; data curation, A.B.; writing—original draft preparation, A.C.; writing—review and editing, M.R.; visualisation, G.P. and L.O.; supervision, A.G. and M.d.V.; project administration, A.P. and L.O.; funding acquisition, A.P. All authors have read and agreed to the published version of the manuscript.

Funding: This research received external funding from the Asylum, Migration, and Integration Fund (FAMI) 2014–2020 of the Italian Ministry of the Interior, and cofinanced by the European Union according to code PROG-2010.

Institutional Review Board Statement: The "PROTECT" project was approved by the Department of Oral and Maxillofacial Sciences, Sapienza, University of Rome (Protocol identifying number: 0000839 on 2 October 2018). The protocol was in accordance with the 1975 Declaration of Helsinki on medical protocols and ethics and its later amendments.

Informed Consent Statement: Informed consent was obtained from all subjects involved in the study.

Data Availability Statement: The datasets used and/or analysed during the current study are available from the corresponding author on reasonable request.

Conflicts of Interest: The authors declare no conflict of interest.

References

1. Lebano, A.; Hamed, S.; Bradby, H.; Gil-Salmeron, A.; Dura-Ferrandis, E.; Garces-Ferrer, J.; Azzedine, F.; Riza, E.; Karnaki, P.; Zota, D.; et al. Migrants' and refugees' health status and healthcare in Europe: A scoping literature review. *BMC Public Health* **2020**, *20*, 1039. [CrossRef] [PubMed]
2. Zenner, D.; Mendez, A.R.; Schillinger, S.; Val, E.; Wickramage, K. Health and illness in migrants and refugees arriving in Europe: Analysis of the electronic Personal Health Record system. *J. Travel Med.* **2022**, *29*, taac035. [CrossRef] [PubMed]
3. Lauritano, D.; Moreo, G.; Carinci, F.; Campanella, V.; Della Vella, F.; Petruzzi, M. Oral Health Status among Migrants from Middle- and Low-Income Countries to Europe: A Systematic Review. *Int. J. Environ. Res. Public Health* **2021**, *18*, 12203. [CrossRef]
4. Pinchi, V.; Focardi, M.; Pradella, F.; Grifoni, R.; Palandri, M.; Norelli, G.A. Day to day issues in the forensic identification practice related to illegal immigration in Italy. *J. Forensic Odonto-Stomatol.* **2017**, *35*, 157–165.

5. Seedat, F.; Hargreaves, S.; Nellums, L.B.; Ouyang, J.; Brown, M.; Friedland, J.S. How effective are approaches to migrant screening for infectious diseases in Europe? A systematic review. *Lancet Infect. Dis.* **2018**, *18*, e259–e271. [CrossRef] [PubMed]
6. Ralli, M.; Marinelli, A.; De-Giorgio, F.; Crescenzi, D.; Vincentiis, M.; Greco, A.; Arcangeli, A.; Ercoli, L. Prevalence of Otolaryngology Diseases in an Urban Homeless Population. *Otolaryngol. Head Neck Surg.* **2022**, *166*, 1022–1027. [CrossRef] [PubMed]
7. Itamura, K.; Hur, K.; Ference, E.; Wrobel, B.; Chambers, T.N. Characterization of Chronic Sinonasal Disease Symptoms in an Urban Homeless Population. *Am. J. Rhinol. Allergy* **2020**, *34*, 494–501. [CrossRef]
8. Lee, S.J.; Thomas, P.; Newnham, H.; Freidin, J.; Smith, C.; Lowthian, J.; Borghmans, F.; Gocentas, R.A.; De Silva, D.; Stafrace, S. Homeless status documentation at a metropolitan hospital emergency department. *Emerg. Med. Australas.* **2019**, *31*, 639–645. [CrossRef]
9. Wadhera, R.K.; Choi, E.; Shen, C.; Yeh, R.W.; Joynt Maddox, K.E. Trends, Causes, and Outcomes of Hospitalizations for Homeless Individuals: A Retrospective Cohort Study. *Med. Care* **2019**, *57*, 21–27. [CrossRef]
10. Westerberg, B.D.; Lango, M.N. Otolaryngology-Related Disorders in Underserved Populations, Otolaryngology Training and Workforce Considerations in North America. *Otolaryngol. Clin. North Am.* **2018**, *51*, 685–695. [CrossRef]
11. Daly, B.; Newton, T.; Batchelor, P.; Jones, K. Oral health care needs and oral health-related quality of life (OHIP-14) in homeless people. *Community Dent. Oral Epidemiol.* **2010**, *38*, 136–144. [CrossRef] [PubMed]
12. Materia, E.; Spadea, T.; Rossi, L.; Cesaroni, G.; Area, M.; Perucci, C.A. Health care inequalities: Hospitalization and socioeconomic position in Rome. *Epidemiol. Prev.* **1999**, *23*, 197–206. [PubMed]
13. Di Stadio, A.; Ralli, M.; Roccamatisi, D.; Scarpa, A.; Della Volpe, A.; Cassandro, C.; Ricci, G.; Greco, A.; Bernitsas, E. Hearing loss and dementia: Radiologic and biomolecular basis of their shared characteristics. A systematic review. *Neurol. Sci.* **2021**, *42*, 579–588. [CrossRef]
14. Pacelli, B.; Zengarini, N.; Broccoli, S.; Caranci, N.; Spadea, T.; Di Girolamo, C.; Cacciani, L.; Petrelli, A.; Ballotari, P.; Cestari, L.; et al. Differences in mortality by immigrant status in Italy. Results of the Italian Network of Longitudinal Metropolitan Studies. *Eur. J. Epidemiol.* **2016**, *31*, 691–701. [CrossRef] [PubMed]
15. Kroening, A.L.H.; Dawson-Hahn, E. Health Considerations for Immigrant and Refugee Children. *Adv. Pediatr.* **2019**, *66*, 87–110. [CrossRef]
16. Ralli, M.; Arcangeli, A.; Ercoli, L. Homelessness and COVID-19: Leaving No One Behind. *Ann. Glob. Health* **2021**, *87*, 11. [CrossRef]
17. Kluge, H.H.P.; Jakab, Z.; Bartovic, J.; D'Anna, V.; Severoni, S. Refugee and migrant health in the COVID-19 response. *Lancet* **2020**, *395*, 1237–1239. [CrossRef]
18. Perri, M.; Dosani, N.; Hwang, S.W. COVID-19 and people experiencing homelessness: Challenges and mitigation strategies. *CMAJ* **2020**, *192*, E716–E719. [CrossRef]
19. Ralli, M.; Cedola, C.; Urbano, S.; Morrone, A.; Ercoli, L. Homeless persons and migrants in precarious housing conditions and COVID-19 pandemic: Peculiarities and prevention strategies. *Eur. Rev. Med. Pharmacol. Sci.* **2020**, *24*, 9765–9767. [CrossRef]
20. Werner, M.T.; Carey, R.M.; Albergotti, W.G.; Lukens, J.N.; Brody, R.M. Impact of the COVID-19 Pandemic on the Management of Head and Neck Malignancies. *Otolaryngol. Head Neck Surg.* **2020**, *162*, 816–817. [CrossRef]
21. Palaia, G.B.M.; Pranno, N.; Ralli, M.; Bruscolini, A.; Capocci, M.; Gaimari, G.; Macali, F.; Corridore, D.; Cotticelli, L.; Covello, F.; et al. Oral Pathologies in Migrants: The Experience of the "PROTECT" Project in 3023 Patients. *Appl. Sci.* **2022**, *12*, 12621. [CrossRef]
22. Karlsmose, B.; Pedersen, H.B.; Lauritzen, T.; Parving, A. Audiometry in general practice: Validation of a pragmatic pure-tone audiometry method. *Scand. Audiol.* **1998**, *27*, 137–142. [CrossRef] [PubMed]
23. Noorhassim, I.; Kaga, K.; Nishimura, K. Pure-tone audiometry and auditory brainstem responses in noise-induced deafness. *Am. J. Otolaryngol.* **1996**, *17*, 31–35. [CrossRef] [PubMed]
24. Fetoni, A.R.; Garzaro, M.; Ralli, M.; Landolfo, V.; Sensini, M.; Pecorari, G.; Mordente, A.; Paludetti, G.; Giordano, C. The monitoring role of otoacoustic emissions and oxidative stress markers in the protective effects of antioxidant administration in noise-exposed subjects: A pilot study. *Med. Sci. Monit.* **2009**, *15*, PR1–PR8.
25. Global Burden of Disease Cancer, C.; Kocarnik, J.M.; Compton, K.; Dean, F.E.; Fu, W.; Gaw, B.L.; Harvey, J.D.; Henrikson, H.J.; Lu, D.; Pennini, A.; et al. Cancer Incidence, Mortality, Years of Life Lost, Years Lived With Disability, and Disability-Adjusted Life Years for 29 Cancer Groups From 2010 to 2019: A Systematic Analysis for the Global Burden of Disease Study 2019. *JAMA Oncol.* **2022**, *8*, 420–444. [CrossRef]
26. Hashim, D.; Genden, E.; Posner, M.; Hashibe, M.; Boffetta, P. Head and neck cancer prevention: From primary prevention to impact of clinicians on reducing burden. *Ann. Oncol.* **2019**, *30*, 744–756. [CrossRef]
27. Syka, J.; Canlon, B. Editorial: Special issue on presbycusis. *Hear. Res.* **2021**, *402*, 108115. [CrossRef]
28. Patel, R.; McKinnon, B.J. Hearing Loss in the Elderly. *Clin. Geriatr. Med.* **2018**, *34*, 163–174. [CrossRef]
29. Osterhammel, D.; Osterhammel, P. High-frequency audiometry. Age and sex variations. *Scand. Audiol.* **1979**, *8*, 73–81. [CrossRef]
30. Riva, M.M.; Cantamessa, F.; Borleri, D.; Mosconi, G. Occupational health and safety of road haulage company employees. *Med. Lav.* **2018**, *109*, 180–189. [CrossRef]
31. McKenna, M.J.; Mills, B.G. Ultrastructural and immunohistochemical evidence of measles virus in active otosclerosis. *Acta Otolaryngol. Suppl.* **1990**, *470*, 130–139; discussion 139–140. [CrossRef] [PubMed]

32. Tatsuo, H.; Ono, N.; Tanaka, K.; Yanagi, Y. SLAM (CDw150) is a cellular receptor for measles virus. *Nature* **2000**, *406*, 893–897. [CrossRef] [PubMed]
33. Arnold, W. Some Remarks on the Histopathology of Otosclerosis. In *Otosclerosis and Stapes Surgery*; Karger Publishers: Basel, Switzerland, 2007; Volume 65, pp. 25–30. [CrossRef]
34. McDermott, J.H.; Molina-Ramirez, L.P.; Bruce, I.A.; Mahaveer, A.; Turner, M.; Miele, G.; Body, R.; Mahood, R.; Ulph, F.; MacLeod, R.; et al. Diagnosing and Preventing Hearing Loss in the Genomic Age. *Trends Hear.* **2019**, *23*, 2331216519878983. [CrossRef] [PubMed]
35. Nelson, D.I.; Nelson, R.Y.; Concha-Barrientos, M.; Fingerhut, M. The global burden of occupational noise-induced hearing loss. *Am. J. Ind. Med.* **2005**, *48*, 446–458. [CrossRef]
36. Frederiksen, T.W.; Ramlau-Hansen, C.H.; Stokholm, Z.A.; Grynderup, M.B.; Hansen, A.M.; Lund, S.P.; Kristiansen, J.; Vestergaard, J.M.; Bonde, J.P.; Kolstad, H.A. Occupational noise exposure, psychosocial working conditions and the risk of tinnitus. *Int. Arch. Occup. Environ. Health* **2017**, *90*, 217–225. [CrossRef]
37. Zhou, X.; Merzenich, M.M. Environmental noise exposure degrades normal listening processes. *Nat. Commun.* **2012**, *3*, 843. [CrossRef]
38. Agrawal, Y.; Platz, E.A.; Niparko, J.K. Risk factors for hearing loss in US adults: Data from the National Health and Nutrition Examination Survey, 1999 to 2002. *Otol. Neurotol.* **2009**, *30*, 139–145. [CrossRef]
39. Rubak, T.; Kock, S.A.; Koefoed-Nielsen, B.; Bonde, J.P.; Kolstad, H.A. The risk of noise-induced hearing loss in the Danish workforce. *Noise Health* **2006**, *8*, 80–87.
40. Sheppard, A.; Ralli, M.; Gilardi, A.; Salvi, R. Occupational Noise: Auditory and Non-Auditory Consequences. *Int. J. Environ. Res. Public Health* **2020**, *17*, 8963. [CrossRef]
41. Hwang, S.W.; Lebow, J.M.; Bierer, M.F.; O'Connell, J.J.; Orav, E.J.; Brennan, T.A. Risk factors for death in homeless adults in Boston. *Arch. Intern. Med.* **1998**, *158*, 1454–1460. [CrossRef]
42. Usatine, R.P.; Gelberg, L.; Smith, M.H.; Lesser, J. Health care for the homeless: A family medicine perspective. *Am. Fam. Physician* **1994**, *49*, 139–146. [PubMed]
43. Ly, T.D.A.; Edouard, S.; Badiaga, S.; Tissot-Dupont, H.; Hoang, V.T.; Pommier de Santi, V.; Brouqui, P.; Raoult, D.; Gautret, P. Epidemiology of respiratory pathogen carriage in the homeless population within two shelters in Marseille, France, 2015–2017: Cross sectional 1-day surveys. *Clin. Microbiol. Infect.* **2019**, *25*, 249.e1–249.e6. [CrossRef] [PubMed]
44. Chaffee, B.W.; Rodrigues, P.H.; Kramer, P.F.; Vitolo, M.R.; Feldens, C.A. Oral health-related quality-of-life scores differ by socioeconomic status and caries experience. *Community Dent. Oral. Epidemiol.* **2017**, *45*, 216–224. [CrossRef] [PubMed]

applied sciences

Article

Patient's Perception and Real Execution of Walking as Physical Exercise: Looking at Self-Efficacy as a Key Variable in Adherence in Patients with Fibromyalgia

Lorena Gutiérrez [1], Ana Myriam Lavín-Pérez [2,3], Patricia Catalá [1], Carmen Écija [1], Daniel Collado-Mateo [2], Alexander Gil-Arias [2] and Cecilia Peñacoba [1,*]

1 Department of Psychology, Rey Juan Carlos University, 28922 Alcorcón, Madrid, Spain
2 GO fitLAB, Ingesport, 28003 Madrid, Spain
3 Centre for Sport Studies, Rey Juan Carlos University, 28942 Fuenlabrada, Madrid, Spain
* Correspondence: cecilia.penacoba@urjc.es; Tel.: +34-91-488-8864

Abstract: Adherence to physical exercise is especially low in patients with fibromyalgia, and there is a tendency to overestimate its performance. This study explores the relationship between perceived and actual walking behavior in patients with fibromyalgia, including self-efficacy as a mediating variable. A prospective study with two time points (with one week between them) was carried out on 132 women diagnosed with fibromyalgia. Self-efficacy towards exercise was assessed at the first time point, and perceived intensity of physical activity and "objective" walking amount (accelerometer) in relation to the previous week were assessed at the second point. Physical and emotional symptoms and the consumption of drugs were also evaluated as covariates. The results show that perception of vigorous-intensity exercise was related to walking (accelerometer) (effort greater than 6 METs). This relationship was explained by the mediating effect of self-efficacy and was not interfered with either by the symptoms or by the consumption of drugs. High self-efficacy regarding physical exercise was directly related to walking behavior despite patients' interpretation of this activity as a vigorous physical exercise. In intervention and rehabilitation programs, it would be interesting to take into account and promote the self-efficacy of patients in relation to physical exercise in order to maintain adherence to walking.

Keywords: fibromyalgia; walking; self-efficacy; perception of walking; accelerometry; mediation analysis

check for updates

Citation: Gutiérrez, L.; Lavín-Pérez, A.M.; Catalá, P.; Écija, C.; Collado-Mateo, D.; Gil-Arias, A.; Peñacoba, C. Patient's Perception and Real Execution of Walking as Physical Exercise: Looking at Self-Efficacy as a Key Variable in Adherence in Patients with Fibromyalgia. *Appl. Sci.* **2023**, *13*, 1191. https://doi.org/10.3390/app13021191

Academic Editors: Massimo Ralli, Gaspare Palaia and Alice Bruscolini

Received: 19 December 2022
Revised: 9 January 2023
Accepted: 12 January 2023
Published: 16 January 2023

1. Introduction

Approximately 3% of the Spanish population suffers from chronic pain problems, with fibromyalgia (FM) being particularly relevant because of its differential characteristics. Fibromyalgia is more common in women than men, with females accounting for 80% of diagnoses nationwide [1]. This somatic syndrome is characterized by the unexplained presence of widespread musculoskeletal pain along with intense fatigue. Both symptoms may be accompanied by additional symptoms such as sleep disturbances, anxiety, depression, or problems with functionality [2]. Regarding data collected worldwide, between 2% and 4% of people with chronic pain are diagnosed with fibromyalgia [3,4], making it an important public health problem in which helping people to successfully cope with their pain or fatigue remains a relevant goal for healthcare professionals and researchers.

For people with FM, a particularly important threat to their quality of life is symptom interference, understood as the perception of how pain or fatigue interfere with the performance of other daily activities, significantly limiting successful adaptation to the symptoms [5]. Therefore, a multidisciplinary approach should not only include pharmacological therapy, but other types of therapeutic resources that help alleviate emotional distress and increase the rewards derived from engaging in meaningful activities versus avoiding activity to reduce pain or other symptoms, such as fatigue [5]. Sedentary behavior has recently emerged as a strong risk factor fin the development of future chronic diseases in healthy individuals reporting low levels of perceived pain [6]. Moreover, after the COVID-19 lockdowns, stress levels have potentially increased sedentary activity globally, which has conditioned the development of rehabilitation programs adapted to the reality faced by patients with chronic pain [7]. Several studies have pointed to the role of physical activity as a resource that can aid in the improvement of pain, fatigue, stiffness, and functional levels, as well as enhance positive emotional states through the reduction of anxiety and depression [8–10]. Multimodal physical exercise interventions have focused on aerobic conditioning, as it can decrease the risk of potentially painful injuries in patients' daily lives [11,12]. Specifically, it appears that moderate-intensity aerobic physical exercise improves pain and fatigue thresholds [9,13,14]. In this context, telerehabilitation has been an innovative scenario in face of the COVID-19 pandemic and has generated similar benefits to face-to-face therapeutic protocols that were implemented before the pandemic [15], providing support to care services for chronic populations. As such, it should be taken into account to improve patients' quality of life [16].

However, despite the reported benefits, there are also potential barriers to adherence to this type of treatment in this population, including the interference of the above-mentioned symptoms. The paradoxical effect of symptoms on physical exercise in patients with chronic pain has been widely reported in the literature, and, although moderate aerobic physical exercise alleviates symptoms in the mid–long term, at the same time said symptoms are the main inhibitor for the initiation and maintenance of physical activity, especially during the initial phases [8,17,18]. Therefore, one explanation for the lack of adherence lies in the difficulties that patients find when reducing their sedentary lifestyle due to an increase of symptoms occurring when physical activity is performed at too high a pace and/or without professional supervision [11,19,20]. Another potential barrier to adherence to physical activity among chronic pain patients may reside in the incompatibility of physical exercise programs with patients' daily routines; in fact, most physical fitness programs require numerous visits to specialized clinics, explaining a dropout rate of approximately 40% [14]. Taking the above into consideration, walking offers greater advantages than other types of high-intensity physical exercise, such as: self-regulation of the pattern itself, low economic cost with minimal interference in normal routines, and relying exclusively on the motivation of the patient [21]. Furthermore, walking appears to be the pattern that most increases perceptions of self-efficacy in terms of managing FM-related problems (e.g., pain, fatigue, or disability). However, considering the sedentary tendencies of the FM population [22,23], it is especially important to recommend initiating walking gradually and regularly [24]. Specifically, based on the recommendations outlined by various chronic pain professionals, walking for at least 60 min in bouts of 20 min with a small rest between bouts, four times a week, over a minimum of six consecutive weeks [25] is recommended, and has been shown to have beneficial effects on long-term health [25–27].

In summary, it is important to analyze walking within the context in which it occurs. This means taking into account patients' symptoms, perceptions about the intensity of the activity, and beliefs regarding their own abilities to walk as a form of physical exercise. In support of this idea, previous research has pointed out the lack of agreement between patients' perceptions and actual exercise performance; because patients with FM often consider light–moderate activities as intense, complete physical recovery becomes impossible, and they abandon the goal [28,29]. Therefore, most clinical research highlights the need for instruments that include both objective and subjective measures of activity, specifically in relation to exercise intensity and capacity [30,31]. One of the most used questionnaires to assess perceptions of activity in FM is the abbreviated version of the International Physical Activity Questionnaire (IPAQ-s) which records time spent performing physical activities as a function of intensity (moderate, light, walking and sitting time). Recent research suggests that the IPAQ-s information must be used to support measures offered by other assessment instruments such as accelerometers [32,33]. The latter type of instrument allows for the monitoring of energy expenditure and directly measures specific behaviors based on intensity, body posture, or counts of the amount of physical activity performed [34]. These types of measurement devices employ metabolic rate measurement units (METs) that attempt to analyze the amount of energy consumed by a person at rest. Thus, 1 MET corresponds to 3.5 mL O_2/kg per minute, which is the minimum amount of oxygen consumption that the body needs to maintain itself. Based on this unit of measurement, the type of physical exercise is differentiated according to its intensity measured in METs: light or gentle exercise has an energy expenditure of 3 METs; light–moderate exercise expends between 3–6 METs and vigorous or intense exercise more than 6 METs [35]. In patients with fibromyalgia, accelerometry is frequently used to assess physical activity in view of the benefits of exercise in this population [10,36,37]. In support of previous research, the first objective of this study has been to compare the measures provided by the IPAQ-s as a subjective measure of the intensity of the activity performed, and the measures provided by the accelerometer whilst walking.

Regarding beliefs about one's own abilities to walk as a form of physical exercise, self-efficacy—understood as the set of all knowledge that people have about their abilities and confidence to achieve a goal or cope with a situation—plays an important role in the physical function of patients suffering from chronic pain [38,39]. In addition, it promotes the positive emotional states necessary to initiate physical exercise [40]. However, the association between perceptions about physical activity and objective walking behavior has not been examined whilst taking into account the effect of beliefs regarding patient abilities and their possible interference with core symptoms such as pain or fatigue or the effects of medication [41]. Therefore, the second main objective of our research aimed to examine the relationships between subjective and objective measures of walking behavior and self-efficacy from a simple mediational model (taking self-efficacy as a mediator variable). To this end, the intention has been to previously control for the possible effects of FM symptoms (i.e., pain, fatigue, sleep or cognitive problems, anxiety, and depression) and medication.

In particular, we hypothesized an overestimation of the intensity of the physical activity performed in comparison to the walking measurement provided by the accelerometers. Likewise, we expected self-efficacy to play a significant mediating role in the performance of walking behavior based on said overestimation of the activity. Fibromyalgia symptoms and drug use were expected to affect the overall model (greater drug use and greater symptoms will have an impact on less walking behavior).

2. Materials and Methods

2.1. Participants and Procedure

The research protocol of the present study was reviewed and approved by the Bioethics Committee (Universidad Rey Juan Carlos, Madrid, Spain, Reference 160520165916, date of approval 9 June 2016). The study followed the ethical guidelines of the Declaration of Helsinki.

Different fibromyalgia associations in the Community of Madrid and Castilla La Mancha (Spain) were contacted. All patients interested in participating had to be women over 18 years of age, be diagnosed with fibromyalgia according to the criteria established by the American College of Rheumatology (ACR) [2], have a prescribed walking pattern, show no impediments to physical activity, and sign the informed consent form. The exclusion criteria that were established to carry out the study were the presence of other comorbid chronic pain diagnoses or severe psychiatric diagnoses. The final participant sample consisted of 132 women.

Regarding data collection in this prospective study, sociodemographic, clinical characteristics (including medication) of the participants and self-efficacy were evaluated first. After the evaluation, an accelerometer was provided to be worn on the wrist for one week. Patients were instructed not to remove the accelerometer except for aquatic activities such as showering or swimming. The final assessment was conducted one week later, when the accelerometers were collected. Perception of physical activity and symptoms (pain, fatigue, unrefreshing sleep, cognitive problems, anxiety and depression) in the previous week were also assessed at this point.

Both self-reports and accelerometer data were collected within the associations. All members of the research team carried out the evaluations. The data were downloaded to a computer and processed with the manufacturer's software.

2.2. *Measures*

2.2.1. First Assessment

Sociodemographic and clinical data. An interview was applied with open-ended questions that the participants had to answer regarding their age, weight, height, marital status, employment status, educational level, and clinical aspects such as the use and type of medication.

Self-efficacy about physical exercise in chronic pain. Self-efficacy related to physical activity was assessed using the Spanish adaptation of the Self-efficacy scale for physical activity scale for chronic pain (SEPAS) [42]. This adaptation reports on the sedentary lifestyle of women with fibromyalgia, physical activity intensity levels and walking pattern recommendations [43] attending to 5 barriers: pain, fatigue, bad weather, feeling stressed, sad, worried, and having a bad day because of fibromyalgia. The SEPAS is composed of 35 items grouped into 3 factors depending on the intensity of the physical exercise performed. For this study, the self-efficacy subscale related to walking while taking advantage of doing other activities was used, which is made up of 5 items that ask about the ability to perform walking related to daily activities despite the aforementioned barriers (e.g., "I feel able to walk at least 30 min, taking advantage of having to do other activities (walk the dog, go to work or shopping) yet (a) I have pain, (b) I have fatigue, (c) There is bad weather, (d) I feel sad, stressed or worried, (e) I have a sick day"). In our sample, the walking self-efficacy subscale showed an internal consistency of 0.85.

2.2.2. Second Assessment

Walking, moderate, and vigorous physical activity (subjective measure). The abbreviated version of the International Physical Questionnaire (IPAQ-s) was used to assess the physical activity performed in the previous week according to the level of intensity (moderate, vigorous intensity, and walking) [44]. Following the instructions, the time measures recorded by the self-report are: days per week, minutes per week, and minutes per day for each type of intensity. Some examples of items are the following: "During the last 7 days, on how many days did you do moderate physical activities such as carrying light weights, bicycling at a regular speed, swimming, or dancing? (Do not include walking)"; and "How much time in total did you usually spend in moderate physical activity on one of those days?" For a correct interpretation of the possible relationships between the intensity levels provided by the IPAQ-s and the intensity levels provided by the accelerometer, the minutes/week results were used for each intensity level.

Walking (objective measure). The ActiGraph wGT3X-BT (Pensacola, FL, USA) was used to record total physical activity in minutes per day and intensity. According to studies in chronic pain populations, episodes of 20 continuous minutes with intensity 0 are excluded from the analysis because it is considered sedentary activity. Time spent walking is considered a moderate-intensity activity and was consequently defined as minutes accumulated between 100 and 759 counts—min-1 [34].

2.2.3. Covariates

Medication. A high percentage of women diagnosed with FM take medication to alleviate symptoms such as pain or fatigue and allow for restful sleep [45,46]. In line with the previous literature, we asked about the number of pills taken per day according to the type of medication: analgesics, sleeping pills, antidepressants and muscle relaxants.

Pain. Pain intensity was measured by the Brief Pain Inventory (BPI) [47]. In particular, the BPI item that refers to the mean intensity of pain experienced during the past week (on a numerical scale ranging from 0—"no pain"—to 10—"greatest pain imaginable"—) was used.

Fatigue, unrefreshing sleep, and cognitive problems. Fatigue intensity, non-refreshing sleep and cognitive problems in the last week were measured by 3 independent items assessing the severity index of symptoms on a 4-point Likert scale (0—"no problem", 3—"severe, persistent, pervasive, with interference in daily activities").

Anxiety and depression symptoms. Anxiety and depressive symptoms were evaluated by the Spanish validation of the Hospital Anxiety and Depression Scale (HADS) [48]. This instrument provides information on the presence of these symptoms in non-psychiatric patients and is made up of 14 items (on a 4-point scale) that are divided (7 items each) into a scale that evaluates anxiety symptoms (e.g., "I feel tense or nervous", "I feel a kind of dread as if something bad is going to happen") and another scale to measure depression symptoms (e.g., "I feel slow and clumsy", "I have lost interest in my personal appearance". In our sample, the internal consistency was 0.70 and 0.72 for the anxiety and depression scale, respectively.

2.3. Data Analysis

The analyses were carried out using SPSS 27.0 (Armonk, NY, USA) and the level of significance was set at $p = 0.05$. The characteristics of the sample were described with frequency distributions, percentages, means, and standard deviations.

First, we explored the possible relationships between the perception of different levels of intensity on physical activity performed by the participants, accelerometer-reported walking behavior, and self-efficacy to perform physical activities using Pearson's correlation analysis.

Finally, we intend to specify the roles of each of the study variables by proposing more complex statistical models. Based on the previous literature, we proposed a simple mediation model in which the subjective measure of the level of intensity (variable X) predicted the time spent walking referred by the accelerometer (variable Y). We considered that this relationship could be influenced by self-efficacy for physical activity (variable M) as can be seen in Figure 1. To test this model, we used model 4 (simple mediation model) of the PROCESS 3.5 macro controlling for the effect of medication, pain, non-restorative sleep, fatigue, cognitive problems, and anxiety and depression symptoms as covariates of the simple mediation model. According to the criteria established by Hayes [49], the variables posed in the model must correlate significantly ($p < 0.05$). The bootstrap technique was used to control the type I error in the sample with a 95% confidence interval. The bootstrap estimates were based on 5000.

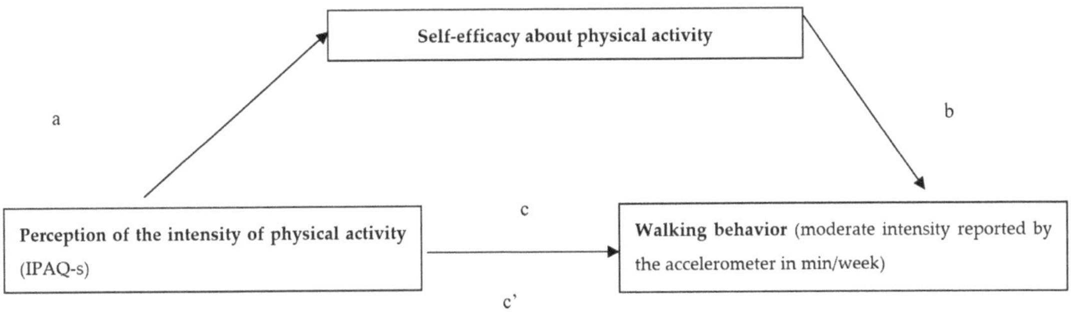

Figure 1. Proposed simple mediation model. Coefficient a is the effect of predictor on outcome; coefficient b is the effect of mediator on outcome; coefficient c is the total effect of predictor on outcome; c′ is the indirect effect that are represented as the effect of predictor on outcome.

3. Results

A total of 158 patients completed the first assessments of the study. However, 132 women completed the subsequent assessments after wearing the accelerometers for one week. Descriptive statistics of sociodemographic and clinical data are presented in Table 1.

Table 1. Sociodemographic and clinical data of participants.

	FM Participants (n = 132)					
Sociodemographic Data	n	(%)	M	(SD)	Max	Min
Age			57.63	10.44	79	32
Marital status						
Married	95	(72.4)				
Divorced or separated	13	(9.8)				
Widow	11	(8)				
Single	13	(9.8)				
Educational level						
Primary studies	89	(66)				
Secondary studies	31	(24.6)				
University studies	12	(9.4)				
Employment status						
Working	30	(22.7)				
Housewife	66	(50)				
Sick leave	15	(11.4)				
Unemployed	21	(15.9)				
Clinical data			*			
Medication						
Analgesics						
Yes	104	(86.7)	4.37	2.54	5	0
No	28	(13.3)				
Sleeping pills						
Yes	77	(65.9)	3.20	3.30	3	0
No	55	(34.1)				
Antidepressants						
Yes	78	(66.7)	3.67	3.31	2	0
No	54	(33.3)				
Muscle relaxants						
Yes	49	(31.9)	1.27	0.55	4	0
No	83	(68.1)				
Others						
Yes	74	(55.4)	1.20	0.41	6	0
No	58	(44.6)				

Note: M (Mean); SD (Standard Deviation); * data collected on the type of medications is number of pills per day.

3.1. *Aim 1: Associations between Subjective Measures of Physical Activity Intensity (IPAQ-s) and Objective Measure of Walking (Accelerometer)*

As shown in Table 2, walking behavior recorded by the accelerometer was positively associated with the perception of activity as vigorous (r = 0.765, p = 0.000). Self-efficacy was positively related to the perception of moderate- (r = 0.324, p = 0.001) and vigorous-intensity (r = 0.258, p = 0.007) IPAQ-s activity and to walking recorded by the accelerometer (r = 0.213, p = 0.026).

In relation to symptoms, there was a negative significant correlation between the perception of moderate activity and depression (r = −0.217, p = 0.013). Similarly, there was a negative significant association between the accelerometer walking measure and both cognitive problems (r = −0.180, p = 0.040) and depression (r = −0.265, p = 0.002). Negative significant relationships were also shown between self-efficacy and pain (r = −0.272, p = 0.004), cognitive problems (r = −0.321, p = 0.001), and depression (r = −0.379, p = 0.000).

Table 2. Pearson correlations.

	Mean	SD	Min–Max	1	2	3	4	5	6	7	8	9	10	11
1. Walking (IPAQ-s)	3587.96 min per day	408.27	0–21,960	–										
2. Moderate activity (IPAQ-s)	131.66 min per day	55.76	0–2940	−0.034	–									
3. Vigorous activity (IPAQ-s)	40.73 min per day	20.61	0–2940	−0.038	0.720 **	–								
4. Walking (Accelerometer)	519.29 min per day	138.71	104.75–1045.75	0.051	0.017	0.765 **	–							
5. Self-efficacy physical activity	4.63 (a.u.)	2.31	0–10	0.156	0.324 **	0.258 **	0.213 *	–						
6. Pain	7.10 (a.u.)	1.83	1–10	0.086	0.082	0.021	0.046	−0.272 **	–					
7. Fatigue	2.29 (a.u.)	0.81	0–3	−0.041	−0.089	−0.058	−0.131	−0.174	0.056	–				
8. Sleep problems	2.60 (a.u.)	0.68	0–3	−0.005	0.018	0.056	−0.049	−0.147	0.261 **	0.185 *	–			
9. Cognitive problems	2.28 (a.u.)	0.82	0–3	0.064	−0.169	−0.084	−0.180 *	−0.321 **	0.250 **	0.260 **	0.150	–		
10. Anxiety	11.86 (a.u.)	3.86	3–21	−0.032	−0.060	−0.006	−0.104	−0.154	0.228 *	0.080	0.087	0.084	–	
11. Depression	9.46 (a.u.)	2.36	3–21	−0.119	−0.217 *	0.099	−0.265 **	−0.379 **	0.175	0.241 **	0.139	0.322 **	0.529 **	–

Note: * $p < 0.05$; ** $p < 0.01$; a.u. arbitrary unit.

Regarding the medication, the only significant relationships were found between analgesics (pills per day) and the perception of exercise as intense (r = 0.258, p = 0.007), and with self-efficacy (r = 0.212, p = 0.025).

3.2. Aim 2: Mediational Role of Self-Efficacy between Perception of Intensity of Physical Activity (Subjective Measure) and Walking Behavior (Objective Measure)

Based on the results obtained in the correlational analyses and according to Hayes' criteria [49] on mediational models, our predictor variable was the perception of intense physical activity, and our outcome variable was walking as a moderate-intensity activity (accelerometers). This finding is consistent with previous studies which indicate an overestimation by patients with fibromyalgia of the intensity of the physical activity that they carry out [28,29]. Based on this finding, we intended to analyze the possible role of self-efficacy in this relationship according to the theoretical model proposed in Figure 1. The proposed simple mediation model resulted in uniformly acceptable model fits (R^2 = 0.286). As hypothesized, perception of physical activity (vigorous) predicted the walking behavior registered by the accelerometer. Specifically, perceptions about the performance of intense physical activity were predictive of walking behavior (c′ = 5.10, t = 1.01, p = 0.008, 95% Confidence Interval (CI) = 0.081, 1.44). In addition, a significant effect of the perception of intense physical activity on self-efficacy was observed (a = 1.63, t = 2.49, p = 0.014, 95% CI = 1.04, 1.34), along with an effect of self-efficacy on walking behavior (b = 4.63, t = 1.70, p = 0.022, 95% CI = 1.93, 5.21). The relationship between the perception of intense physical activity and walking behavior was explained by the mediating effect of self-efficacy (c = 3.16, t = 3.38, p = 0.004, 95% CI = 2.85, 4.72). Of all the symptoms considered as covariates in the model, only pain (B = −0.31, t = 0.13, p = 0.014, 95% CI = −0.54, −0.064) and depressive symptoms (B = −0.19, t = −2.99, p = 0.003, 95% CI = −0.31, −0.065) influenced the effect of perception of physical activity on self-efficacy but did not explain the total effect of the mediation model. Regarding the type of medication, only analgesics (pills per day) had a mediating effect between the perception of intense physical activity and self-efficacy (B = −0.20, t = −2.23, p = 0.027, 95% CI = −0.37, −0.02). In other words, despite the main symptoms of fibromyalgia and the use of analgesics, the perception of carrying out an intense physical activity influenced objective walking behavior through the perception of self-efficacy (see Figure 2).

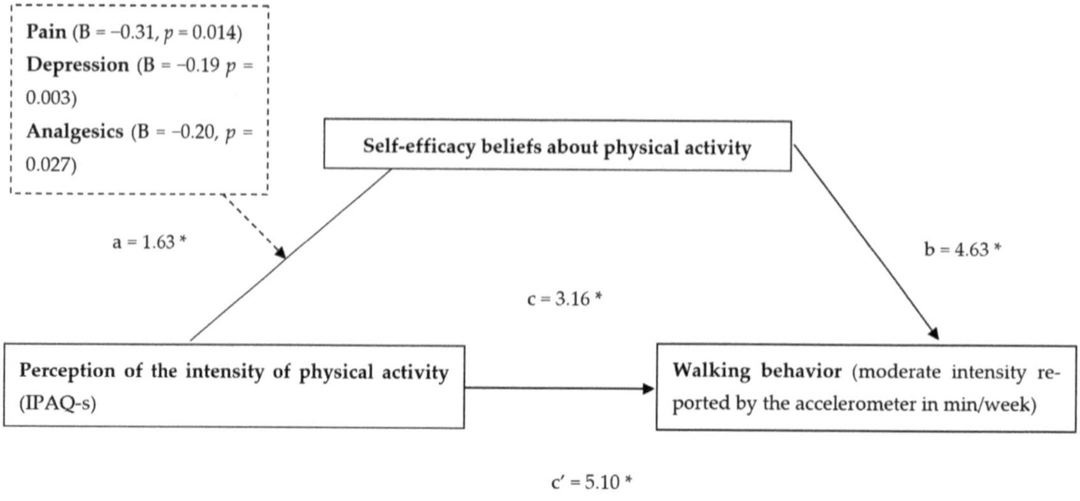

Figure 2. Paths between perception of physical activity (subjective measure) and walking behavior (objective measure) through self-efficacy beliefs about physical activity, controlling for symptoms and medication. Note * $p < 0.05$.

4. Discussion

In this study we used a simple mediational model to clarify the role of perceptions regarding walking held by women with fibromyalgia and the role they play in adhering to a pattern recommended as a way to decrease adverse symptoms. We found that although participants viewed walking as an intense activity, it was positively associated with objective walking through the influence of beliefs regarding their abilities about physical exercise (self-efficacy). Overall, the hypothesized model paths were supported, and model fits were uniformly acceptable. Based on the stability of the results found, the clinical implications of the present study are discussed below.

The first step we undertook was to explore possible mismatches between the subjects' perceptions of walking and the objective measures of their walking pattern. Our results clearly suggest a significant association between subjective measures of vigorous activity and the objective measure of walking. Although previous studies have reached the same conclusions with smaller fibromyalgia or chronic pain samples [50–53], our interest was also focused on determining which psychological and physical factors may explain the differences between subjectivity and observable performance in patients' walking behavior. On this topic, the previous literature on fibromyalgia shows that the discordance between assessment measures can be explained by problems in central pain processing. This fosters the persistent occurrence of chronic widespread pain, among other common symptoms of fibromyalgia such as fatigue and cognitive problems [54,55] along with the presence of anxious and/or depressive symptoms [28,29]. Therefore, we set out to explore the possible effect of physical and emotional symptoms on the association between the subjective (self-reported) and the objective (accelerometer-based) measures of walking.

The correlation results showed significant relationships between the measurements provided by the IPAQ-s questionnaire and the accelerometer. Specifically, of the two types of intensity provided by the self-report, it was the perception of intense activity that was significantly associated with walking, considering that the latter is a moderate-intensity activity following the indications of Gusi et al. [25] regarding physical exercise guidelines in fibromyalgia. These data suggest the suitability of both instruments for evaluating the physical activity of chronic pain patients, since it is not only the amount of exercise performed that is important, but the subjective vision that patients have about exercise as well. In this context, the perception of the intensity of a given activity could have a negative or positive impact on adherence to this guideline as part of the recommended treatment for fibromyalgia.

In view of the above, the following question may be asked: what factors influence women with fibromyalgia to perceive moderate activity as intense? According to the contributions offered by Jones et al. [31], the physical symptoms typical of the disease influence the patients' subjective evaluation of physical exercise. Even memory and attention problems have an impact on patients not remembering accurately the type of exercise they have performed [56], and therefore self-report measurements may be biased. Our results show that depressive symptoms correlate both with perceived moderate-intensity physical activity (IPAQ-s) and with the objective measure of walking (accelerometers), while in the latter case (accelerometers) the association with cognitive problems must also be included. Based on these preliminary analyses, the model we suggest considers these symptoms as possible factors influencing the predictive effect of intensity perception on objective walking behavior.

In line with our hypothesis, the perception of vigorous intensity in the activity predicted the walking pattern (accelerometers) carried out for a week. Despite the presence of adverse symptoms, these do not seem to explain the lack of activity in the participants as claimed by the fear-of-movement model of Vlaeyen et al. [57]. The central focus of this model lies in catastrophic beliefs about pain, which exacerbate pain intensity which, in turn, leads to avoidance of physical activity and thus contributes to greater disability [22,58,59]. Although the literature on the effects of fatigue is scarcer, this symptom has also been found to promote avoidance behaviors in women suffering from FM in previous studies [60,61].

Our results show that only pain and depression influenced the association between the perception of intensity and self-efficacy, but nevertheless there was no influence on the model as a whole (the model considers the association between the perception of intensity and walking—accelerometers—through self-efficacy); therefore, within the aforementioned model, the symptoms do not seem to have an influence on lower physical activity. Another factor that is indirectly associated with a decrease in time devoted to physical activity is the use of antidepressants, which consequently increases sedentary activity. In a study of people with severe mental illness, it was found that people medicated with antidepressants (SSRIs) are more sedentary than those who do not take them [62,63]. Our results indicate that only analgesics (and not antidepressants) negatively influenced the relationship between the perception of activity as intense and self-efficacy; however, they did not affect the model proposed in its entirety and therefore walking performance. It should also be noted that, globally, the lack of participation in physical activities is due to multiple factors, especially environmental influences. For example, the scarcity of parks, pedestrian paths, lack of sports and leisure facilities, or the use of cell phones correlate positively with a sedentary lifestyle [64,65]. Again, the environment in which people develop seems to have a strong influence on the decisions they make regarding exercise, so this is an issue that also needs to be resolved in studies dedicated to people with chronic pain.

Our results are supported by the findings offered by new affective-motivational models on pain [66,67]. These theories take context as a reference to understand that the relationship between catastrophic beliefs and activity avoidance is not direct, but rather, several factors come into play that show a preference for hedonic goals—aimed at avoiding the threat of symptoms—or for achievement goals when performing physical activity [68,69]. In this sense, it is cognitive and emotional factors that may mostly explain behavior beyond the effect of physical symptoms [29,70]. Many patients lack the confidence to exercise despite their symptoms and lack adaptive beliefs that exercise will help them manage their disease and improve their quality of life [71,72]. Along these lines, self-efficacy turns out to be an essential component for undertaking a new activity, as it provides the basis of motivation to start an exercise. Consistent with previous research that has studied the role of self-efficacy within the motivational context of chronic pain, our findings suggest that beliefs about ability regarding physical exercise play an important mediating role between patients' perception and the actual execution of walking. This is despite perceiving it as an activity of greater intensity, and taking into account pain, as a primary symptom of FM, depressive symptoms, and drug use (analgesics). Previous studies have also confirmed the mediation of self-efficacy in the results obtained in the management of pain catastrophizing and perceived pain control thanks to cognitive behavioral therapy [73]. In addition, self-efficacy, fibromyalgia tender points, perceived impact of illness, physical functioning, as well as reported long-term anxiety and depression, have all also been shown to be important predictors of objective pain [74,75]. Therefore, our findings suggest that feeling able to perform any physical exercise would dampen the effect of perceiving walking as intense and would help to successfully cope with fibromyalgia.

Regarding the potential clinical applications of the present study, we believe that focusing solely on reducing pain intensity would be an inadequate therapeutic goal. From a rehabilitative perspective, the most effective treatments for fibromyalgia are those that combine cognitive-behavioral therapy with graded physical exercise for each patient profile [9,54]. Following these guidelines, we consider it appropriate to work on promoting self-efficacy through motivational and volitional interventions that encourage a preference for achievement goals over hedonic goals based on pain/movement avoidance. Furthermore, our data tentatively suggest that a physical program is also effective with individuals in whom large differences exist between perceived intensity of physical exercise and target performance. This clinical approach makes sense if fear of injury is adequately addressed as a belief easily modifiable by ability; the minimization of such beliefs are also related to positive emotional states and to less symptom interference [76,77].

Despite its possible implications, there are limitations to this study that warrant some attention. First, it might be expected that the female volunteers who participated in our study felt more confident in performing physical exercise compared to the general population with fibromyalgia. Therefore, it would be interesting for future studies to also control for the role of catastrophizing as a possible behavioral inhibitor, as pointed out by the previous literature [59,78–80]. Another limitation is regarding the age of the participants, as the mean age was older than in other studies. A well-known age-associated health problem is cognitive decline, which we assessed using a single question with responses distributed on a Likert-type scale. However, it is possible that a single self-report item is not a valid measure to detect the appearance of certain cognitive problems that could be influencing the proposed mediational model. For this reason, it is important to control for this health problem as a possible covariate, and to try to assess it with other appropriate instruments. Furthermore, it seems that the discordance between subjective and objective measures about physical exercise is higher in women than in men with fibromyalgia [51,81], which in turn limits the generalization of the results obtained. Finally, our results indicate that the proposed model is not affected by the symptoms. However, symptoms such as fatigue have several dimensions that explain why this experience is unique to each patient [82]. Therefore, it would be important to assess physical fatigue and mental fatigue independently to explore their effect on this model. Along this line, we have included the effect of medication on the execution of physical activity and the perception of its intensity. Finally, as we have noted, women with FM tend to perceive greater activity and greater intensity than the exercise they have done. However, we do not know whether this result is confirmed in the healthy population. Therefore, we propose to resolve this question in future studies by replicating this model with the inclusion of a control group.

In summary, previous research on chronic pain from motivational perspectives such as the goals-conflict theory [69], emphasize the role of adaptive cognitions to take into account during the multidisciplinary treatment of fibromyalgia. In this context, our results highlight the role of self-perceived capacity beliefs (self-efficacy), through adherence to walking, in mitigating patients' dysfunctionality within the rehabilitative context.

Author Contributions: Conceptualization, L.G., P.C., C.É., A.M.L.-P., D.C.-M., A.G.-A. and C.P.; methodology, L.G., P.C. and C.P.; formal analysis, L.G. and C.P.; investigation, L.G., P.C., C.É., A.M.L.-P., D.C.-M., A.G.-A. and C.P.; resources, L.G., P.C., C.É., A.M.L.-P., D.C.-M., A.G.-A. and C.P.; data curation, L.G., P.C., C.É. and C.P.; writing—original draft preparation, L.G. and C.P.; writing—review and editing, L.G., P.C., C.É. and C.P.; visualization, L.G., C.É. and P.C.; supervision, C.P.; project administration, C.P. All authors have read and agreed to the published version of the manuscript.

Funding: This work was funded by the Health Research Found (Fondo de Investigación en Salud). grant number PI17/00858 from the Instituto de Salud Carlos III (Spain), co-financed by the European Union through the Fondo Europeo de Desarrollo Regional (FEDER). Moreover, the author A.M.L.-P. is supported by the Industrial Doctorate Spanish National grant program, part of the Strategic Plan on Science and Innovation Support of the Spanish Ministry of Science, Innovation and Universities (predoctoral industry grant).

Institutional Review Board Statement: All procedures performed in studies involving human participants were in accordance with the ethical standards of the institutional and/or national research committee and with the 1964 Helsinki Declaration and its later amendments or comparable ethical standards. The study was approved by the Bioethics Committee of Rey Juan Carlos University (Reference PI17/00858).

Informed Consent Statement: Informed consent was obtained from all subjects involved in the study.

Data Availability Statement: The data presented in this study are available on request from the corresponding author.

Conflicts of Interest: The authors declare no conflict of interest. The funders had no role in the design of the study, in the collection, analyses or interpretation of data; in the writing of the manuscript; or in the decision to publish the results.

References

1. Ministerio de Sanidad, Política Social e Igualdad. Fibromialgia. 2011. Available online: https://www.sanidad.gob.es/profesionales/prestacionesSanitarias/publicaciones/docs/fibromialgia.pdf (accessed on 12 September 2022).
2. Wolfe, F.; Clauw, D.J.; Fitzcharles, M.A.; Goldenberg, D.L.; Katz, R.S.; Mease, P.; Russell, A.S.; Russell, I.J.; Winfield, J.B.; Yunus, M.B. The American College of Rheumatology preliminary diagnostic criteria for fibromyalgia and measurement of symptom severity. *Arthr. Care Res.* **2010**, *62*, 600–610. [CrossRef] [PubMed]
3. Kaltsas, G.; Tsiveriotis, K. Fibromyalgia. In *Endotext*; Feingold, K.R., Anawalt, B., Boyce, A., Chrousos, G., de Herder, W.W., Dhatariya, K., Dungan, K., Hershman, J.M., Hofland, J., Kalra, S., et al., Eds.; MDText.com, Inc.: South Dartmouth, MA, USA, 2020.
4. Häuser, W.; Fitzcharles, M.A. Facts and myths pertaining to fibromyalgia. *Dialogues Clin. Neurosci.* **2018**, *20*, 53–62. [CrossRef] [PubMed]
5. Steiner, J.L.; Bigatti, S.M.; Ang, D.C. Trajectory of change in pain, depression and physical functioning after physical activity adoption in fibromyalgia. *J. Health Psychol.* **2015**, *20*, 931–941. [CrossRef] [PubMed]
6. Marques, A.; Santos, T.; Martins, J.; Matos, M.G.; Valeiro, M.G. The association between physical activity and chronic diseases in European adults. *Eur. J. Sport Sci.* **2018**, *18*, 140–149. [CrossRef] [PubMed]
7. Iannuccelli, C.; Lucchino, B.; Gioia, C.; Dolcini, G.; Favretti, M.; Franculli, D.; Di Franco, M. Mental health and well-being during the COVID-19 pandemic: Stress vulnerability, resilience and mood disturbances in fibromyalgia and rheumatoid arthritis. *Clin. Exp. Rheumatol.* **2021**, *39* (Suppl. 130), 153–160. [CrossRef]
8. Andrade, A.; Steffens, R.d.A.K.; Vilarino, G.T.; Sieczkowska, S.M.; Coimbra, D.R. Does volume of physical exercise have an effect on depression in patients with fibromyalgia? *J. Affect. Disord.* **2017**, *208*, 214–217. [CrossRef]
9. Häuser, W.; Ablin, J.; Fitzcharles, M.A.; Littlejohn, G.; Luciano, J.V.; Usui, C.; Walitt, B. Fibromyalgia. Nature reviews. *Dis. Prim.* **2015**, *1*, 15022.
10. Vural, M.; Berkol, T.D.; Erdogdu, Z.; Pekedis, K.; Kuçukserat, B.; Aksoy, C. Evaluation of the effectiveness of an aerobic exercise program and the personality characteristics of patients with fibromyalgia syndrome: A pilot study. *J. Phys. Ther. Sci.* **2014**, *26*, 1561–1565. [CrossRef]
11. Bircan, C.; Karasel, S.A.; Akgün, B.; El, O.; Alper, S. Effects of muscle strengthening versus aerobic exercise program in fibromyalgia. *Rheumat. Intern.* **2008**, *28*, 527–532. [CrossRef]
12. Williams, D.A.; Clauw, D.J. Understanding fibromyalgia: Lessons from the broader pain research community. *J. Pain* **2009**, *10*, 777–791. [CrossRef]
13. Busch, A.J.; Webber, S.C.; Brachaniec, M.; Bidonde, J.; Bello-Haas, V.D.; Danyliw, A.D.; Schachter, C.L. Exercise therapy for fibromyalgia. *Curr. Pain Head Rep.* **2011**, *15*, 358–367. [CrossRef]
14. Meiworm, L.; Jakob, E.; Walker, U.A.; Peter, H.H.; Keul, J. Patients with fibromyalgia benefit from aerobic endurance exercise. *Clin. Rheum.* **2000**, *19*, 253–257. [CrossRef]
15. Paolucci, T.; de Sire, A.; Ferrillo, M.; di Fabio, D.; Molluso, A.; Patruno, A.; Pesce, M.; Lai, C.; Ciacchella, C.; Saggino, A.; et al. Telerehabilitation proposal of mind-body technique for physical and psychological outcomes in patients with fibromyalgia. *Front. Physiol.* **2022**, *13*, 917956. [CrossRef]
16. Hernando-Garijo, I.; Ceballos-Laita, L.; Mingo-Gómez, M.T.; Medrano-de-la-Fuente, R.; Estébanez-de-Miguel, E.; Martínez-Pérez, M.N.; Jiménez-del-Barrio, S. Immediate effects of a telerehabilitation program based on aerobic exercise in women with fibromyalgia. *Int. J. Environ. Res. Public Health* **2021**, *18*, 2075. [CrossRef]
17. Masquelier, E.; D'haeyere, J. Physical activity in the treatment of fibromyalgia. *Jt. Bone Spine* **2021**, *88*, 105202. [CrossRef]
18. Collado-Mateo, D.; Lavín-Pérez, A.M.; Peñacoba, C.; Del Coso, J.; Leyton-Román, M.; Luque-Casado, A.; Amado-Alonso, D. Key factors associated with adherence to physical exercise in patients with chronic diseases and older adults: An umbrella review. *Int. J. Environ. Res Public Health* **2021**, *18*, 2023. [CrossRef]
19. López-Roig, S.; Pastor, M.A.; Peñacoba, C. Prevalence and predictors of unsupervised walking and physical activity in a community population of women with fibromyalgia. *Rheum. Intern.* **2016**, *36*, 1127–1133. [CrossRef]
20. Turk, D.C.; Vierck, C.J.; Scarbrough, E.; Crofford, L.J.; Rudin, N.J. Fibromyalgia: Combining pharmacological and nonpharmacological approaches to treating the person, not just the pain. *J. Pain* **2008**, *9*, 99–104. [CrossRef]
21. Rooks, D.; Gautam, S.; Romeling, M.; Cross, M.; Stratigakis, D.; Evans, B.; Katz, J. Group exercise, education, and combination self-management in women with fibromyalgia: A randomized trial. *Arch. Intern. Med.* **2007**, *167*, 2192–2200. [CrossRef]
22. Ellingson, L.D.; Stegner, A.J.; Schwabacher, I.J.; Lindheimer, J.B.; Cook, D.B. Catastrophizing Interferes with Cognitive Modulation of Pain in Women with Fibromyalgia. *Pain Med.* **2018**, *19*, 2408–2422. [CrossRef]
23. McLoughlin, M.J.; Colbert, L.H.; Stegner, A.J.; Cook, D.B. Are women with fibromyalgia less physically active than healthy women? *Med. Sci. Sport. Exerc.* **2011**, *43*, 905–912. [CrossRef] [PubMed]
24. Sitthipornvorakul, E.; Klinsophon, T.; Sihawong, R.; Janwantanakul, P. The effects of walking intervention in patients with chronic low back pain: A meta-analysis of randomized controlled trials. *Muscul. Sci. Pract.* **2018**, *34*, 38–46. [CrossRef] [PubMed]
25. Gusi, N.; Parraca, J.; Adsuar; Olivares, P. Physical exercise and Fibromyalgia. In *Physical Exercise Guidelines for People with Fibromyalgia*; Penacho, A., Rivera, J., Pastor, M.A., Gusi, N., Eds.; Asociación Divulgación Fibromialgia: Vitoria, Spain, 2009; pp. 39–56, ISBN 978-84-692-2868-5.
26. Pastor, M.A.; López-Roig, S.; Sanz, Y.; Peñacoba, C.; Cigarán, M.; Velasco, L.; Lledó, A.; Écija, C. Walking as physical exercise in Fibromyalgia: An elicitation study from the Theory of Planned Behavior. *An. Psicol.* **2015**, *31*, 433–446. [CrossRef]

27. Sanz-Baños, Y.; Pastor, M.Á.; Velasco, L.; López-Roig, S.; Peñacoba, C.; Lledo, A.; Rodríguez, C. To walk or not to walk: Insights from a qualitative description study with women suffering from fibromyalgia. *Rheum. Int.* **2016**, *36*, 1135–1143. [CrossRef] [PubMed]
28. Doerr, J.M.; Fischer, S.; Nater, U.M.; Strahler, J. Influence of stress systems and physical activity on different dimensions of fatigue in female fibromyalgia patients. *J. Psychosom. Res.* **2017**, *93*, 55–61. [CrossRef]
29. Soriano-Maldonado, A.; Amris, K.; Ortega, F.B.; Segura-Jiménez, V.; Estévez-López, F.; Álvarez-Gallardo, I.C.; Aparicio, V.A.; Delgado-Fernández, M.; Henriksen, M.; Ruiz, J.R. Association of different levels of depressive symptoms with symptomatology, overall disease severity, and quality of life in women with fibromyalgia. *Qual Life Res.* **2015**, *24*, 2951–2957. [CrossRef]
30. Jones, K.D.; Burckhardt, C.S.; Clark, S.R.; Bennett, R.M.; Potempa, K.M. A randomized controlled trial of muscle strengthening versus flexibility training in fibromyalgia. *J. Rheum.* **2002**, *29*, 1041–1048.
31. Jones, K.D.; Burckhardt, C.S.; Bennett, J.A. Motivational interviewing may encourage exercise in persons with fibromyalgia by enhancing self-efficacy. *Arthritis Rheum.* **2004**, *51*, 864–867. [CrossRef]
32. Benítez-Porres, J.; Delgado, M.; Ruiz, J.R. Comparison of physical activity estimates using International Physical Activity Questionnaire (IPAQ) and accelerometry in fibromyalgia patients: The Al-Andalus study. *J. Sport. Sci.* **2013**, *31*, 1741–1752. [CrossRef]
33. Segura-Jiménez, V.; Álvarez-Gallardo, I.C.; Romero-Zurita, A.; Camiletti-Moirón, D.; Munguía-Izquierdo, D.; Carbonell-Baeza, A.; Ruiz, J.R. Comparison of physical activity using questionnaires (leisure time physical activity instrument and physical activity at home and work instrument) and accelerometry in fibromyalgia patients: The al-Ándalus project. *Arch. Phys. Med. Rehab.* **2014**, *95*, 1903–1911.e2. [CrossRef]
34. Matthews, C.E.; Hagströmer, M.; Pober, D.M.; Bowles, H.R. Best practices for using physical activity monitors in population-based research. *Med. Sci. Sport. Exerc.* **2012**, *44* (Suppl. 1), S68–S76. [CrossRef]
35. Jetté, M.; Sidney, K.; Blümchen, G. Metabolic equivalents (METS) in exercise testing, exercise prescription, and evaluation of functional capacity. *Clin. Cardiol.* **1990**, *13*, 555–565. [CrossRef]
36. Häuser, W.; Klose, P.; Langhorst, J.; Moradi, B.; Steinbach, M.; Schiltenwolf, M.; Busch, A. Efficacy of different types of aerobic exercise in fibromyalgia syndrome: A systematic review and meta-analysis of randomised controlled trials. *Arthr. Res. Ther.* **2010**, *12*, R79. [CrossRef]
37. Kelley, G.A.; Kelley, K.S.; Hootman, J.M.; Jones, D.L. Exercise and global well-being in community-dwelling adults with fibromyalgia: A systematic review with meta-analysis. *BMC Public Health* **2010**, *10*, 198. [CrossRef]
38. Pastor, M.A.; Lledó, A.; Pons, N.; López-Roig, S. Dolor y fibromialgia. Aportaciones desde la Psicología. In *Fibromialgia y Promoción de la Salud. Herramientas de Intervención Psicosocial*; Peñacoba, C., Ed.; Dykinson: Madrid, Spain, 2012; pp. 77–122.
39. Van Liew, C.; Brown, K.C.; Cronan, T.A.; Bigatti, S.M. The effects of self-efficacy on depression and pain in fibromyalgia syndrome: Does initial depression matter? *J. Muscul. Pain* **2013**, *21*, 113–125. [CrossRef]
40. Kamping, S.; Bomba, I.C.; Kanske, P.; Diesch, E.; Flor, H. Deficient modulation of pain by a positive emotional context in fibromyalgia patients. *Pain* **2013**, *154*, 1846–1855. [CrossRef]
41. Sánchez, A.I.; Martínez, M.P.; Miró, E.; Medina, A. Predictors of the pain perception and self-efficacy for pain control in patients with fibromyalgia. *Span. J. Psychol.* **2011**, *14*, 366–373. [CrossRef]
42. López-Roig, S.; Pastor-Mira, M.Á.; Núñez, R.; Nardi, A.; Ivorra, S.; León, E.; Peñacoba, C. Assessing Self-Efficacy for Physical Activity and Walking Exercise in Women with Fibromyalgia. *Pain Manag. Nurs.* **2021**, *22*, 571–578. [CrossRef]
43. Pastor, M.Á.; López-Roig, S.; Lledó, A.; Peñacoba, C.; Velasco, L.; Schweiger-Gallo, I.; Cigarán, M.; Ecija, C.; Limón, R.; Sanz, Y. Combining motivational and volitional strategies to promote unsupervised walking in patients with fibromyalgia: Study protocol for a randomized controlled trial. *Trials* **2014**, *15*, 120. [CrossRef]
44. Ainsworth, B.E.; Haskell, W.L.; Herrmann, S.D.; Meckes, N.; Bassett, D.R.; Tudor-Locke, C., Jr.; Greer, J.L.; Vezina, J.; Whitt-Glover, M.C.; Leon, A.S. 2011 Compendium of Physical Activities: A second update of codes and MET values. *Med. Sci. Sport. Exerc.* **2011**, *43*, 1575–1581. [CrossRef]
45. Häuser, W.; Bernardy, K.; Arnold, B.; Offenbächer, M.; Schiltenwolf, M. Efficacy of multicomponent treatment in fibromyalgia syndrome: A meta-analysis of randomized controlled clinical trials. *Arthritis Rheum.* **2009**, *61*, 216–224. [CrossRef] [PubMed]
46. Martín, J.; Torre, F.; Aguirre, U.; González, N.; Padierna, A.; Matellanes, B.; Quintana, J.M. Evaluation of the interdisciplinary PSYMEPHY treatment on patients with fibromyalgia: A randomized control trial. *Pain Med.* **2014**, *15*, 682–691. [CrossRef] [PubMed]
47. Cleeland, C.S.; Ryan, K.M. Pain assessment: Global use of the Brief Pain Inventory. *Ann. Acad. Med.* **1994**, *23*, 129–138.
48. Terol, M.C.; López-Roig, S.; Rodríguez-Marín, J.; Martín-Aragón, M.; Pastor, M.A.; Reig, M.T. Propiedades psicométricas de la Escala Hospitalaria de Ansiedad y Estrés en población española. *Ansiedad Estrés* **2007**, *13*, 163–176.
49. Hayes, A.F. *Introduction to Mediation, Moderation, and Conditional Process Analysis: A Regression-Based Approach*; Guilford Press: New York, NY, USA, 2013.
50. Bandak, E.; Amris, K.; Bliddal, H.; Danneskiold-Samsøe, B.; Henriksen, M. Muscle fatigue in fibromyalgia is in the brain, not in the muscles: A case-control study of perceived versus objective muscle fatigue. *Ann. Rheum. Dis.* **2013**, *72*, 963–966. [CrossRef]

51. Estévez-López, F.; Álvarez-Gallardo, I.C.; Segura-Jiménez, V.; Soriano-Maldonado, A.; Borges-Cosic, M.; Pulido-Martos, M.; Aparicio, V.A.; Carbonell-Baeza, A.; Delgado-Fernández, M.; Geenen, R. The discordance between subjectively and objectively measured physical function in women with fibromyalgia: Association with catastrophizing and self-efficacy cognitions. The al-Ándalus project. *Disab. Rehab.* **2018**, *40*, 329–337. [CrossRef]
52. Hidding, A.; van Santen, M.; De Klerk, E.; Gielen, X.; Boers, M.; Geenen, R.; Vlaeyen, J.; Kester, A.; van der Linden, S. Comparison between self-report measures and clinical observations of functional disability in ankylosing spondylitis, rheumatoid arthritis and fibromyalgia. *J. Rheum.* **1994**, *21*, 818–823.
53. Kop, W.J.; Lyden, A.; Berlin, A.A.; Ambrose, K.; Olsen, C.; Gracely, R.H.; Williams, D.A.; Clauw, D.J. Ambulatory monitoring of physical activity and symptoms in fibromyalgia and chronic fatigue syndrome. *Arthritis Rheum.* **2005**, *52*, 296–303. [CrossRef]
54. Clauw, D.J. Fibromyalgia: A clinical review. *JAMA* **2004**, *311*, 1547–1555. [CrossRef]
55. Woolf, C.J. Central sensitization: Implications for the diagnosis and treatment of pain. *Pain* **2011**, *152* (Suppl. 3), S2–S15. [CrossRef]
56. Etnier, J.L.; Karper, W.B.; Gapin, J.I.; Barella, L.A.; Chang, Y.K.; Murphy, K.J. Exercise, fibromyalgia, and fibrofog: A pilot study. *J. Phys. Act. Health* **2009**, *6*, 239–246. [CrossRef]
57. Vlaeyen, J.W.S.; Kole-Snijders, A.M.J.; Boeren, R.G.B.; van Eek, H. Fear of movement/(re)injury in chronic low back pain and its relation to behavioral performance. *Pain* **1995**, *62*, 363–372. [CrossRef]
58. Crombez, G.; Van Damme, S.; Eccleston, C. Hypervigilance to pain: An experimental and clinical analysis. *Pain* **2005**, *116*, 4–7. [CrossRef]
59. Galvez-Sánchez, C.M.; Montoro, C.I.; Duschek, S.; Del Paso, G. Pain catastrophizing mediates the negative influence of pain and trait-anxiety on health-related quality of life in fibromyalgia. *Qual. Life Res.* **2020**, *29*, 1871–1881. [CrossRef]
60. Peñacoba, C.; López-Gómez, I.; Pastor-Mira, M.A.; López-Roig, S.; Ecija, C. Contextualizing goal preferences in fear-avoidance models. Looking at fatigue as a disabling symptom in fibromyalgia patients. *PLoS ONE* **2021**, *16*, e0254200. [CrossRef]
61. Velasco, L.; López-Gómez, I.; Gutiérrez, L.; Écija, C.; Catalá, P.; Peñacoba, C. Exploring the Preference for Fatigue-avoidance Goals as a Mediator Between Pain Catastrophizing, Functional Impairment, and Walking Behavior in Women with Fibromyalgia. *Clin. J. Pain* **2021**, *38*, 182–188. [CrossRef]
62. Carvalho, A.F.; Sharma, M.S.; Brunoni, A.R.; Vieta, E.; Fava, G.A. The Safety, Tolerability and Risks Associated with the Use of Newer Generation Antidepressant Drugs: A Critical Review of the Literature. *Psychother. Psychosom.* **2016**, *85*, 270–288. [CrossRef]
63. Serretti, A.; Mandelli, L. Antidepressants and body weight: A comprehensive review and meta-analysis. *J. Clin. Psychiatry* **2010**, *71*, 1259–1272. [CrossRef]
64. Fennell, C.; Barkley, J.E.; Lepp, A. The relationship between cell phone use, physical activity, and sedentary behavior in adults aged 18–80. *Comput. Hum. Behav.* **2019**, *90*, 53–59. [CrossRef]
65. Park, J.H.; Moon, J.H.; Kim, H.J.; Kong, M.H.; Oh, Y.H. Sedentary Lifestyle: Overview of Updated Evidence of Potential Health Risks. *Korean J. Fam. Med.* **2020**, *41*, 365–373. [CrossRef]
66. Crombez, G.; Eccleston, C.; Van Damme, S.; Vlaeyen, J.W.; Karoly, P. Fear-avoidance model of chronic pain: The next generation. *Clin. J. Pain* **2012**, *28*, 475–483. [CrossRef] [PubMed]
67. Vervoort, T.; Trost, Z. Examining Affective-Motivational Dynamics and Behavioral Implications Within the Interpersonal Context of Pain. *J. Pain* **2017**, *18*, 1174–1183. [CrossRef] [PubMed]
68. Karsdorp, P.A.; Ranson, S.; Schrooten, M.; Vlaeyen, J. Pain catastrophizing, threat, and the informational value of mood: Task persistence during a painful finger pressing task. *Pain* **2012**, *153*, 1410–1417. [CrossRef] [PubMed]
69. Van Damme, S.; Van Ryckeghem, D.M.L.; Wyffels, F.; Van Hulle, L.; Crombez, G. No pain no gain? Pursuing a competing goal inhibits avoidance behavior. *Pain* **2012**, *153*, 800–804. [CrossRef] [PubMed]
70. Taylor, A.M.; Phillips, K.; Patel, K.V.; Turk, D.C.; Dworkin, R.H.; Beaton, D.; Clauw, D.J.; Gignac, M.; Markman, J.D.; Williams, D.A.; et al. Assessment of physical function and participation in chronic pain clinical trials: IMMPACT/OMERACT recommendations. *Pain* **2016**, *157*, 1836–1850. [CrossRef]
71. Peñacoba, C.; Pastor, M.Á.; López-Roig, S.; Velasco, L.; Lledo, A. Walking Beliefs in Women with Fibromyalgia: Clinical Profile and Impact on Walking Behavior. *Clin. Nurs. Res.* **2017**, *26*, 632–650. [CrossRef]
72. Slade, S.C.; Patel, S.; Underwood, M.; Keating, J.L. What are patient beliefs and perceptions about exercise for nonspecific chronic low back pain? A systematic review of qualitative studies. *Clin. J. Pain* **2014**, *30*, 995–1005. [CrossRef]
73. Turner, J.A.; Holtzman, S.; Mancl, L. Mediators, moderators, and predictors of therapeutic change in cognitive-behavioral therapy for chronic pain. *Pain* **2007**, *127*, 276–286. [CrossRef]
74. Van liew, C.; Brown, K.C.; Cronan, T.A.; Bigatti, S.M.; Kothari, D.J. Predictors of pain and functioning over time in fibromyalgia syndrome: An autoregressive path analysis. *Arthr. Care Res.* **2013**, *65*, 251–256. [CrossRef]
75. Sahar, K.; Thomas, S.A.; Clarke, S.P. Adjustment to fibromyalgia: The role of domain-specific self-efficacy and acceptance. *Aust. J. Psychol.* **2016**, *68*, 29–37. [CrossRef]
76. Busch, A.J.; Webber, S.C.; Richards, R.S.; Bidonde, J.; Schachter, C.L.; Schafer, L.A.; Danyliw, A.; Sawant, A.; Dal Bello-Haas, V.; Rader, T.; et al. Resistance exercise training for fibromyalgia. *Cochrane Syst. Rev.* **2013**, *2013*, CD010884. [CrossRef]
77. O'Connor, S.R.; Tully, M.A.; Ryan, B.; Bleakley, C.M.; Baxter, G.D.; Bradley, J.M.; McDonough, S.M. Walking exercise for chronic musculoskeletal pain: Systematic review and meta-analysis. *Arch. Phys. Med. Rehab.* **2015**, *96*, 724–734.e3. [CrossRef]
78. Eccleston, C.; Crombez, G. Worry and chronic pain: A misdirected problem-solving model. *Pain* **2007**, *132*, 233–236. [CrossRef]

79. Sanz-Baños, Y.; Pastor-Mira, M.Á.; Lledó, A.; López-Roig, S.; Peñacoba, C.; Sánchez-Meca, J. Do women with fibromyalgia adhere to walking for exercise programs to improve their health? Systematic review and meta-analysis. *Disab. Rehab.* **2018**, *40*, 2475–2487. [CrossRef]
80. Sullivan, M.J.; Thorn, B.; Haythornthwaite, J.A.; Keefe, F.; Martin, M.; Bradley, L.A.; Lefebvre, J.C. Theoretical perspectives on the relation between catastrophizing and pain. *Clin. J. Pain* **2001**, *17*, 52–64. [CrossRef]
81. Lindström Egholm, C.; Krogh, N.S.; Pincus, T.; Dreyer, L.; Ellingsen, T.; Glintborg, B.; Kowalski, M.R.; Lorenzen, T.; Madsen, O.R.; Nordin, H.; et al. Discordance of Global Assessments by Patient and Physician Is Higher in Female than in Male Patients Regardless of the Physician's Sex: Data on Patients with Rheumatoid Arthritis, Axial Spondyloarthritis, and Psoriatic Arthritis from the DANBIO Registry. *J. Rheum.* **2015**, *42*, 1781–1785. [CrossRef]
82. Velasco-Furlong, L.; Gutiérrez-Hermoso, L.; Mateos-Pintado, B.; Castro, D.G.; Blanco-Rico, S.; Sanromán-Canelada, L.; López-Roig, S.; Pastor-Mira, M.Á.; Peñacoba-Puente, C. The 4 U's Rule of Fibromyalgia: A Proposed Model for Fatigue in a Sample of Women with Fibromyalgia: A Qualitative Study. *Intern. J. Environ. Res. Public Health* **2020**, *17*, 6224. [CrossRef]

applied sciences

Review

Motor Outcome Measures in Pediatric Patients with Congenital Muscular Dystrophies: A Scoping Review

Ilaria Cavallina [1], Rossella D'Alessandro [1], Chiara Brusa [2], Elisa Panero [1,3], Enrica Rolle [1], Francesca Rossi [4], Tiziana Mongini [2] and Federica Silvia Ricci [1,*]

[1] Section of Child and Adolescent Neuropsychiatry, Department of Public Health and Pediatric Sciences, University of Turin, 10126 Turin, Italy
[2] Division of Neurology, Department of Neuroscience, University of Turin, 10126 Turin, Italy
[3] Department of Mechanical and Aerospace Engineering, Politecnico di Torino, 10129 Turin, Italy
[4] Rehabilitation Service, Public Health and Pediatric Sciences Department, A.O.U. Città della Salute e della Scienza-Regina Margherita Children's Hospital, 10126 Turin, Italy
* Correspondence: federica.ricci@unito.it

Abstract: Congenital muscular dystrophies (CMDs) are a heterogeneous group of genetic neuromuscular disorders. They usually occur at birth or in early childhood, with delayed acquisition of motor milestones, and diffuse muscle weakness. A dystrophic pattern is evident on the muscle biopsy. They are highly variable both in terms of severity and clinical evolution and in terms of pathogenetic biochemical mechanisms. The aim of this review is to collect and summarize the current knowledge of motor function in pediatric patients with congenital muscular dystrophies and the instruments used to assess it. This scoping review was conducted using the methodology of PRISMA (extension for Scoping Reviews, PRISMA-ScR). Two databases were queried from January 2002 to November 2022. Articles were identified based on title and abstract. Full-text papers published in peer-reviewed English-language journals were selected. It emerged that motor functional aspects are still underinvestigated in CMD patients, probably due to the rarity of these conditions and the phenotypic variability. The scales used to assess motor function are heterogeneous, as are the age groups considered. Finally, the predominant type of research design is cross-sectional; few studies analyze the progression of motor function over time. All these factors make it difficult to correlate the results of different publications and stress the need for more accurate and shared protocols for assessing motor function in these patients.

Keywords: congenital muscular dystrophies; motor outcome measures; motor function; natural history; motor assessment

check for updates

Citation: Cavallina, I.; D'Alessandro, R.; Brusa, C.; Panero, E.; Rolle, E.; Rossi, F.; Mongini, T.; Ricci, F.S. Motor Outcome Measures in Pediatric Patients with Congenital Muscular Dystrophies: A Scoping Review. *Appl. Sci.* **2023**, *13*, 1204. https://doi.org/10.3390/app13021204

Academic Editors: Massimo Ralli, Gaspare Palaia and Alice Bruscolini

Received: 29 December 2022
Revised: 12 January 2023
Accepted: 13 January 2023
Published: 16 January 2023

1. Introduction

Congenital muscular dystrophies are a heterogeneous group of rare neuromuscular disorders. The overall prevalence of CMD is estimated at between 0.6–0.9 cases per 100,000 births. By definition, onset is usually at birth or in early childhood [1]. Suspicion of CMD in later life may be due to the presence of mild or previously overlooked symptoms [2].

Common symptoms of the various forms of CMD are early muscle weakness, hypotonia, and joint contractures. These features may be more or less prominent. The typical floppy infant presentation may be observed in severe cases; in milder cases, antigravity movements of the limbs may be preserved, with greater impairment at the axial level. Joint contractures may sometimes be associated with ligamentous laxity and may be more or less diffuse.

In the advanced stages of the disease, cardiac and respiratory failure may occur. Brain involvement may also be present, usually manifesting as an intellectual disability [2].

Differential diagnosis in children with suspected congenital muscular dystrophy or congenital myopathy is determined by clinical features, age of onset, and inheritance pattern [3].

Congenital muscular dystrophies are characterized by dystrophic components on muscle biopsy [4]. However, heterogeneity in findings at biopsy has blurred the distinctions between disorders once thought to be distinct.

Today, with improved sequencing technology, genetic testing is the gold standard, as it speeds up the diagnostic process and reduces costs [3]. Thirty-seven genes are currently associated with congenital muscular dystrophy phenotypes. The most common types are collagen VI-related disorders (12–19%), dystroglycanopathies (12–25%), laminin-alpha2-related dystrophies (10–37%), and selenoprotein N (11.65%). Fukuyama CMD is the most prevalent form of CMD in Japan [5].

Although advances in research have improved the understanding of the disease mechanisms, there is still no commonly accepted approach to the treatment and care of CMD patients. The causes could be attributed to the rarity of the diseases and the great variability of clinical phenotypes.

In this scoping review, the current knowledge of the natural history and trends of motor function in pediatric patients with congenital muscular dystrophy is collected and summarized. In addition, the assessment tools used to describe motor function according to the form or stage of the disease are also described.

Collecting and sharing data on disease trajectories using longitudinal studies could add significant information to what is currently available. Therefore, we aim to make the scientific community aware of the lack of data and the future efforts needed to increase evidence, also considering possible innovative drug treatments.

2. Materials and Methods

2.1. Protocol and Registration

The PRISMA methodology for scoping reviews was employed to conduct the review. The results are presented following the Preferred Reporting Items for Systematic reviews and Meta-Analyses extension for Scoping Reviews (PRISMA-ScR) Checklist [6].

No a priori protocol was registered. Additional information about the process can be obtained from the corresponding author on request.

2.2. Inclusion Criteria

The inclusion criteria were categorized according to the scheme Population, Concept, and Context (PCC) recommended by the Joanna Briggs Institute for scoping reviews [7].

2.3. Population

We searched for articles including descriptions of motor function in patients with congenital muscular dystrophies (CMD), including alpha-dystroglycanopathies, Emery–Dreifuss muscular dystrophies, collagen VI-related dystrophy, and Merosin-deficient congenital muscular dystrophies. For this review, we did not consider non-congenital dystrophies and other neuromuscular disorders. We included articles that focused on pediatric populations.

2.4. Concept

We selected articles that analyzed and described motor function and its progression over time.

2.5. Context

No cultural, geographical, race, or gender-specific limits were considered for our review.

2.6. Information Sources

We considered articles available from 2002 to November 2022. Selected keywords were combined to create search strategies, adjusted for each screened database.

Articles were searched in the following databases: PubMed/MEDLINE, Embase.

2.7. Search Methods

We used a combination of Medical Subject Headings (MeSH) and free text (where terms were not present in the MeSH Database) to search the entries.

In consultation with a librarian from the Federated Library of Medicine (BFM) at the University of Turin, we developed a working framework for a search strategy (see Appendix A).

The Appendix A show the search process (search strings) used to retrieve the final articles from PubMed/MEDLINE and Embase. References from relevant articles were searched for the inclusion of additional papers not previously identified through the systematic search.

2.8. Level of Evidence and Qualitative Analysis of Eligible Articles

The quality of the eligible articles was assessed according to the levels of evidence established by the JBI. The qualitative assessment was performed by two authors (I.C. and F.R.) independently. All selected papers are case series so the level of evidence is 4b.

3. Results

A total of 16 articles focusing on motor outcome measures in pediatric patients with congenital muscular dystrophies were included in the review after selection.

There were specific articles on collagen VI-related dystrophies (COL6-RD) (n = 2), Fukuyama congenital muscular dystrophies (FCMD) (n = 2), laminopathies or LMNA-related congenital muscular dystrophies (LMNA-RD) (n = 1), LAMA2-related (Merosin-deficient) congenital muscular dystrophies (LAMA2-RD) (n = 2), SEPN1-related myopathies (SEPN1-RM) (n = 2), and a-dystroglycanopathies–FKRP mutations (FKRP) (n = 1), as well as other articles in which different pathologies were analyzed (n = 6).

3.1. Selection of Sources of Evidence

The articles were examined by two authors (I.C. and F.R.), and eligibility for inclusion was determined independently; in the case of discordant opinions between the reviewers, the eligibility of the article was discussed until consensus was reached.

Articles were initially screened based on titles and abstracts according to the population, concept, and context elements previously described. Duplicates were identified and removed. Only full-text papers published in peer-reviewed journals and English were selected. A total of 2690 articles were found: 912 in PubMed/MEDLINE, 1788 in Embase (see Figure 1). The flowchart of the selection of evidence is shown.

3.2. Synthesis of Results

3.2.1. Studies in Patients with COL6-RD Diseases

The data were grouped according to the type of congenital muscular dystrophy. Articles including different forms were grouped in a separate table. For each article, the author, year of publication, type of congenital muscular dystrophy, instruments used for the evaluation, and results are reported.

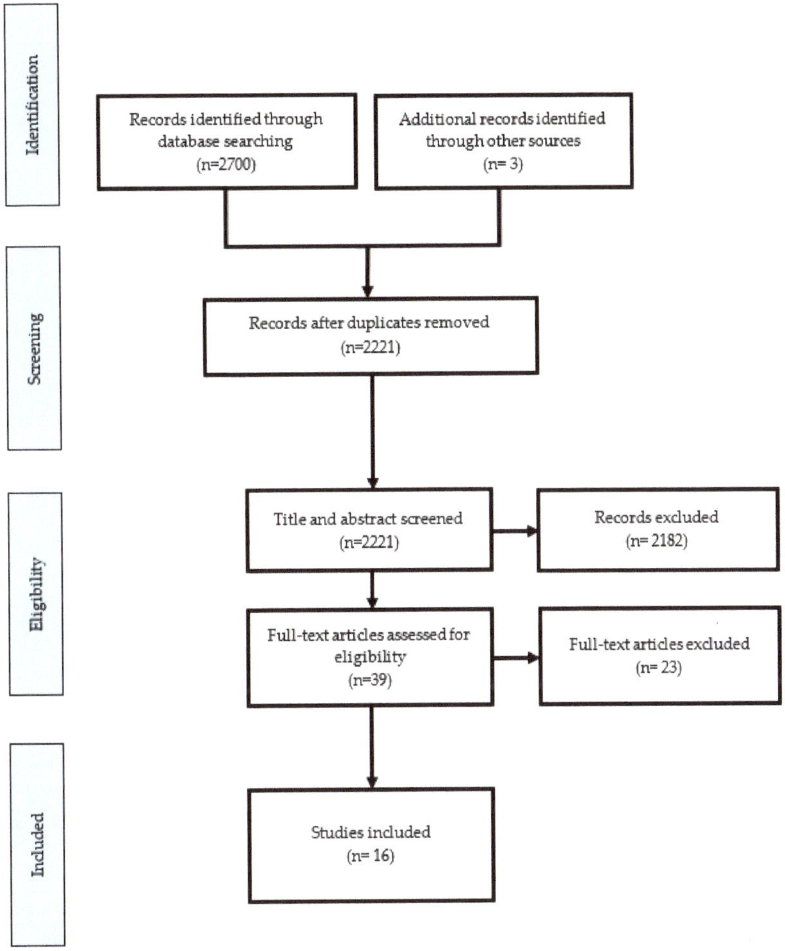

Figure 1. Flowchart of selection of sources of evidence.

We found two articles regarding motor outcomes in patients with COLVI-RD. The details of each article are reported in Table 1.

Table 1. Motor outcome measures in patients with COL6-RD (UCMD: Ullrich congenital muscular dystrophy, BM: Bethlem myopathy; COL6-RD: collagen VI-related dystrophies; MFM32: Motor Function Measure; 6 MWT: 6 min walk test; NSAA: North Star Ambulatory Assessment; LoA: Loss of Ambulation).

Authors, Year	Study Design	Type of CMD	Sample (Size, Sex, Age)	Assessment	Results
Yonekawa et al., 2013 [8]	Case series Cross-sectional	UCMD	n = 33; 18 F/15 M; 11 ± 6.6 years	Clinical assessment	Head control, sitting, and independent ambulation were completed at median ages of 4 months, 9 months, and 18 months, respectively. **Independent sitting and walking**: 81% (25/31). LoA: 8.8 ± 2.9 years (n = 11).

Table 1. *Cont.*

Authors, Year	Study Design	Type of CMD	Sample (Size, Sex, Age)	Assessment	Results
Natera-de et al., 2021 [9]	Case series Cross-sectional	COL6-RD	n = 119; 57 F/62 M; 5–73 years	MFM32, 6MWT, NSAA, timed function tests.	
		UCMD	n = 38; 5–34 years		**MFM32**: 44.17 (±17.4). **NSAA**: 1.8 (±4.4). **6MWT**: 27.5 (±72.6). LoA 22/30—range of ages 4–15 years; 8/38 never achieved independent ambulation.
		Intermediate COL6-RD	n = 35; 6–48 years		**MFM32**: 62.47 (±15.17). **NSAA**: 10.7 (±8.9). **6MWT**: 257.2 (±169.8). LoA 8/35—range of ages 6–12 years.
		BM	n = 46; 5–73 years		**MFM32**: 79.72 (±12.23). **NSAA**: 21.9 (±8.4). **6MWT**: 442.3 (±106.5). Range of ages at LoA (y): all ambulant.

3.2.2. Studies in Patients with LMNA-RD

We found one article regarding motor outcomes in patients with LMNA-RD. The details of the article are reported in Table 2.

Table 2. Motor outcome measures in patients with LMNA-RD (EDMD: Emery–Dreifuss muscular dystrophy, LoA: Loss of Ambulation).

Authors, Year	Study Design	Type of CMD	Sample (Size, Sex, Age)	Assessment	Results
Fan et al., 2021 [10]	Case series Cross-sectional	LMNA-RD mutations	n = 84	Clinical assessment	
		LMNA-related congenital muscular dystrophy	n = 41; 6.5 (1.5–16.0) years		**Independent walking**: 14 subjects. LoA: 7 patients (4.5 years; 3–6 years). **Spinal deformities**: 23 (56.1%). **Contractures**: 21 (51.2%).
		EDMD	n = 32; 10.6 (1.5–43) years		**Independent walking**: all subjects. **Spinal deformities**: 17 (53.1%). **Contractures**: 24 (75.0%).

3.2.3. Studies in Patients with LAMA2—CMD

We found two articles regarding motor outcomes in patients with LAMA2-CMD. The details of the articles are reported in Table 3.

Table 3. Motor outcome measures in patients with LAMA2-RD (KAFO: Knee–Ankle–Foot Orthosis).

Authors, Year	Study Design	Type of CMD	Sample (Size, Sex, Age)	Assessment	Results
Zambon et al., 2020 [11]	Case Series Longitudinal	LAMA2-RD	n = 46; 25 F/21 M; The median age at first and last assessment was 2 years and 12.1 years, respectively (range 0.1–14.8 and 12.1–22.7 years)	Clinical assessment	Analysis of longitudinal data demonstrated a linear yearly rate of increase of 6.6° for right elbow flexion and 3.1° for knee flexion contractures.
		LAMA2-RD with complete laminin alpha 2 deficiency (CD)	n = 42		Motor delay: 42/42 subjects. Independent sitting (in months): mean 13.6, median 12 (range 6–40). Independent walking: 1 subject. Walking with support ± KAFO: 5 subjects. Standing with support: 3 subjects. Scoliosis (median age of onset, years; range): 36 subjects (6.3; 1.9–14.8). Scoliosis surgery (median age at surgery, years; range): 9 subjects (11.6; 7.7–13.5). **Contractures**—Right elbow flexion contractures >30° were observed in 26/34 (76%) subjects at a median age of 6.7 years and >60° in 18/34 (5%) subjects at a median age of 8.7 years. Long finger flexor contractures were reported in 30/42 (71%) CD subjects.
		LAMA2-RD with partial laminin alpha 2 deficiency (PD)	n = 4		Motor delay: 4/4 Independent sitting (in months): mean 10.3, median 10.1 (8.1–13.1). Independent walking: 2 subjects. Walking with support ± KAFO: 2 subjects. Standing with support: 0 Scoliosis: 1 subject. Scoliosis surgery: 0. **Contractures**—No subjects had right elbow contractures >30°. Long finger flexor contractures were reported in 2/4 (50%) subjects.
Tan et al., 2021 [12]	Case Series Cross-sectional	LAMA2-CMD	n = 116; 44 F/72 M; Age of last follow-up: 6.4 (0.3–27.3) years	Clinical assessment	**Head control**: 76.3% (87/114)—(65 achieving after 4 months old) at a median age of 6.0 months (2.0–36.0 months). **Independent sitting**: 92.6% (100/107)—(51 achieving after 10 months old) at a median age of 11.0 months (6.0–36.0 months). **Independent ambulation over 1.5 years old**: 18.4% (18/98)—(14 achieving after 18 months old) at a median age of 27.0 months (18.0–84.0 months). **Regression of motor function**: 31.2% (34/109)—Head control (n = 7), rolling (n = 9), independent sitting (n = 17), and ambulation (n = 9) at median (range) ages of 9.8 (6.8–11.0), 6.0 (3.8–12.0), 8.0 (4.1–19.0), and 8.0 (1.7–11.0) years, respectively. **Spinal deformity**: 48% (54/111). Scoliosis occurred in 40.5% (45/111) at a median age of 6.0 years (0.5–12.0 years) and lordosis occurred in 8.1% (9/111) at a median age of 3.0 years (2.0–7.0 years). **Contractures** in 109 patients involved the knees at first, then the ankles, elbows, and hips in sequence. They progressed rapidly during ages 6–9 with rates of 82.6%, 73.9%, 69.6%, and 43.5%, respectively.

3.2.4. Studies in Patients with FCMD

We found two articles regarding motor outcomes in patients with FCMD. The details of the articles are reported in Table 4.

Table 4. Motor outcome measures in patients with FCMD (FCMD: Fukuyama; GMFM: Gross Motor Function Measure; HMFS: Hammersmith Motor Function).

Authors, Year	Study Design	Type of CMD	Sample (Size, Sex, Age)	Assessment	Results
Harada et al., 2022 [13]	Case series Cross-sectional	Fukuyama—FCMD	n = 13; 5 F/8 M; 1–17 years (mean 9.6)	The Upper Limb Function Measure of Muscular Dystrophy, GMFM, HFMS	The group of patients scoring ≤5 on the Upper Limb Function Measure of Muscular Dystrophy showed significantly lower HMFS ($p = 0.018$) and GMFM ($p = 0.020$) scores than patients scoring ≥6.
Ishigaki et al., 2018 [14]	Case series Cross-sectional	Fukuyama—FCMD	n = 207; 103 F/104 M; 0–42 years (mean 6)	Clinical assessment	**Phenotype**—severe 79 (38%), typical 102 (49%), mild 22 (11%), unknown 4 (2%). **Maximum motor development** Severe—without head control: 37 subjects (18%); with head control: 42 subjects (20%). Typical—sitting without support: 49 subjects (24%); sliding on the buttocks: 53 subjects (26%). Mild—crawling: 6 subjects (3%); walking with support: 5 subjects (2%); walking independently: 5 subjects (2%); climbing stairs: 6 subjects (3%). **Maximum motor development over age 5 years (N = 114)** Severe—without head control: 8 (7.0%); with head control: 12 (11%). Typical—sitting without support: 23 (20%); sliding on the buttocks: 46 (40%). Mild—crawling: 14 (12%); walking with support: 3 (2.6%); walking independently: 4 (3.5%); climbing stairs: 4 (3.5%).

3.2.5. Studies in Patients with SEPN1-RM

We found two articles regarding motor outcomes in patients with SEPN1-RM. The details of the articles are reported in Table 5.

Table 5. Motor outcome measures in patients with SEPN1-RM (SPN1-RM: SEPN1-related myopathies HFMS: Hammersmith Motor Function).

Authors, Year	Study Design	Type of CMD	Sample (Size, Sex, Age)	Assessment	Results
Silwal et al., 2020 [15]	Case series Cross-sectional	SPN1-RM	n = 60; 31 F/29 M. Age range at last assessment: 2–58 years (median 15 years, mean 16.5 years)	Clinical assessment	**Independent walking**: 50/60 patients (83%) (median age at last evaluation: 14 years; range 2.5 to 36 years). Four patients were only able to take steps indoors by holding furniture or walking with assistance. LoA: 8/50. **Scoliosis**: 45/60 (75%). Median age at onset of scoliosis: 12.1 years.
	Case series Longitudinal		n = 25; 13 F/12 M. Age range at last assessment: 2.5–24 years (median 14 years)	HFMS	**HFMS**: estimated annual change was −0.55; of timed 10 m walk, 0.16 s; of timed rise from floor sitting, 0.86 s; and of timed rise from floor lying, 0.87 s. **Spinal stiffness**: 21/25 (84%) at a median age of 10 years (range 2.0–15.6 years).
Villar-Quiles et al., 2020 [16]	Case series Cross-sectional	SPN1-RM	n = 132; 50,8% F/49.2% M; 2–58 years (mean 18.2 ± 11.8)	Clinical assessment	**Delayed motor milestones**: 79/97 (81.4%). Poor head control: 56/97 (57.7%). Delayed gait acquisition: 31/97 (32%). **Spinal stiffness**: 86/98 (87,8%). **Scoliosis**: 87/101 (86,1%). **Contractures**: 64.4% of cases; Achilles tendon (57.4%), hip flexors (50%), elbows (35.2%), or knees (31.5%).

3.2.6. Studies in Patients with FKRP Mutations

We found one article regarding motor outcomes in patients with FKRP mutations. The details of the article are reported in Table 6.

Table 6. Motor outcome measures in patients with FKRP mutations (6MWT: 6 min walk test; MHFMS: Modified Hammersmith Motor Function; PEDI: Pediatric Evaluation of Disability Inventory; PUL: Performance of Upper Limb; TFTs: timed function tests).

Authors, Year	Study Design	Type of CMD	Sample (Size, Sex, Age)	Assessment	Results
Gedlinske et al., 2020 [17]	Case series Longitudinal	FKRP mutations	n = 30; 10 F/20 M. Age at last visit: 14.1 years (11.8–18.1)	Myometry, 6MWT, MHFMS, PEDI, PUL, TFTs.	Mean annual change in tests of motor function: **Strength test**: elbow flexion 0.80 (0.08 to 1.52); elbow extension 0.86 (0.31 to 1.42); shoulder abduction 0.47 (0.04 to 0.91); knee extension 0.78 (−0.84 to 2.40); knee flexion 1.23 (−0.19 to 2.66); hand grip 3.29 (1.23 to 5.35). **6MWT**: −23.41 m (−40.50 to -6.31). **MHFMS**: −0.65 (−0.90 to −0.39). **PEDI**: −1.33 (−3.47 to 0.08). **PUL**: -0.55 (−1.36 to 0.25). **Timed function tests**: 10 m walking speed −0.16 m/s (−0.22 to −0.10); climbing stairs −0.13 steps/s (−0.18 to 0.07); arising from supine position −0.03 rise/s (−0.04 to −0.02).
		Homozygous C.826 > A	n = 6; 2 F/4 M. Age at last visit: 17.3 years (14.5–18.1)		Mean annual change in tests of motor function: **Strength test**: elbow flexion 2.77 (1.66 to 3.88); elbow extension 2.16 (1.32 to 3.00); shoulder abduction 1.24 (0.58 to 1.90); knee extension 5.27 (3.16 to 7.37); knee flexion 3.60 (1.52 to 5.69). **6MWT**: 19.03 m (0.21 to 37.86). **MHFMS**: −0.11 (−0.57 to 0.35). **PEDI**: 2.79 (0.12 to 5.45). **PUL**: −0.21 (−4.34 to 3.92). **Timed function tests**: 10 m walking speed −0.03 m/s (−0.11 to 0.05); climbing stairs −0.01 steps/s (−0.09 to 0.07); arising from supine position −0.01 rise/s (−0.03 to 0.00).
		Other FKRP genotypes	n = 24; 8 F/16 M. Age at last visit: 13.7 years (11.2–17.9)		Mean annual change in tests of motor function: **Strength test**: elbow flexion 0.16 (−0.48 to 0.80); elbow extension 0.39 (−0.12 to 0.91); shoulder abduction 0.12 (−0.32 to 0.55); knee extension −0.69 (−1.92 to 0.53); knee flexion 0.28 (−1.12 to 1.67). **6MWT**: −35.53 m (−47.50 to −23.66). **MHFMS**: −0.90 (−1.18 to −0.62). **PEDI**: −2.94 (−4.26 to −1.62). **PUL**: −0.84 (−2.48 to 0.79). **Timed function tests**: 10 m walking speed −0.20 m/s (−0.25 to −0.15); climbing stairs −0.17 steps/s (−0.22 to −0.12); arising from supine position −0.04 rise/s (−0.05 to −0.03).

3.2.7. Studies in Patients with Various Forms of CMD

We found six articles regarding motor outcomes in patients with various forms of CMD. The details of the articles are reported in Table 7.

Table 7. *Cont.*

Authors, Year	Study Design	Type of CMD	Sample (Size, Sex, Age)	Assessment	Results
Vuillerot et al., 2014 [21]	Case series Cross-sectional	Total CMD	n = 191; 83 F/86 M; 14.8 ± 11.0 years	MFM32	Ambulant 56.8% **MFM32**: total 60.5 ± 24,8. Standing position and transfers (D1): 37.2 ± 32.3; axial and proximal motor function (D2): 73.1 ± 27.1; distal (D3): 81.6 ± 20.4.
		LAMA2-RD	n = 43; 20 F/16 M; 12.2 ± 7.7 years.		Ambulant 48% **MFM32**: total 49.7 ± 28.3. Standing position and transfers (D1): 21.3 ± 29.9; axial and proximal motor function (D2): 62.1 ± 32.9; distal (D3): 71.1 ± 26.2.
		COL6-RD	n = 100; 48 F/46 M; 16.1 ± 12.0 years.		Ambulant 58.3% **MFM32**: total 64.6 ± 21.7. Standing position and transfers (D1): 41.4 ± 31.5; axial and proximal motor function (D2): 78.6 ± 23.1; distal (D3): 87.1 ± 15.2.
		Abnormal glycosylation of dystroglycan	n = 8; 4 F/4 M; 12.8 ± 6.6 years.		Ambulant 62.5% **MFM32**: total 65.6 ± 32.6. Standing position and transfers (D1): 54.2 ± 35.1; axial and proximal motor function (D2): 74 ± 34,6; distal (D3): 72.6 ± 30.4.
		Other CMD	n = 40; 16 F/22 M; 17.4 ± 11.3 years.		Ambulant 57.9% **MFM32**: total 61.8 ± 23.8. Standing position and transfers (D1): 39.6 ± 31.7; axial and proximal motor function (D2): 72.8 ± 26.2; distal (D3): 84.1 ± 17.6.
Le Goff et al., 2021 [22]	Case series Longitudinal	COL6-RD	n = 23; 11 F/12 M; 9 (7.0–13.5) years.	MFM32	Ambulant 70% **MFM32**: total 69 (59.5–77.0). Standing position and transfers (D1): 16 (7.0–24.0); axial and proximal motor function (D2): 32 (29.5–34.0); distal (D3): 20 (18.5–20.0). Longitudinal estimates of average change per year: −4.05 (in ambulant patients −3.38; in non-ambulant −5.28). Standing position and transfers (D1): −1.99; axial and proximal motor function (D2): −1.40; distal (D3): −0.74.
		LAMA2-RD	n = 21; 11 F/10 M; 7 (5.0–9.0) years.		Ambulant 19% **MFM32:** total 44 (27.0–52.0). Standing position and transfers (D1): 1 (1.0–4.0); axial and proximal motor function (D2): 29 (15.0–31.0); distal (D3): 14 (13.0–16.0). Longitudinal estimates of average change per year: −2.62 (in ambulant patients −2.02; in non-ambulant −2.69). Standing position and transfers (D1): −0.72; axial and proximal motor function (D2): -1.59; distal (D3): −0.27.

Table 7. *Cont.*

Authors, Year	Study Design	Type of CMD	Sample (Size, Sex, Age)	Assessment	Results
Meilleur et al., 2015 [23]	Case series Cross-sectional		n = 22 (year 1); 11 F/11 M; 10.1 (5.8-21.2) years. n = 32 (year 2); 19 F/13 M; 9.7 (4.8-22.2) years. n = 33 (years 1 and 2)	MFM32, QUEST, Goniometry, Myometry, 6MWT, HFMS, NSAA, TFTs	Ambulant 36% (12/33) Year 1 **MFM32** (21/22): total 58.9 (25.1–93.9). Standing position and transfers (D1): 31.8 (0–87.6).; axial and proximal motor function (D2): 71.5 (21.9–98.2); distal (D3): 84.8 (49.2–100.0). **Myometry** (21/22): elbow flexion R: 6.8 (1.6–11.9); elbow flexion L: 6.9 (1.6–10.9); elbow extension R: 4.1 (1.2–8.0); elbow extension L: 4.6 (1.0–10.1); knee flexion R: 11.7 (2.7–20.5); knee flexion L: 12.2 (1.4–22.9); knee extension R: 12.9 (2.0–25.4); knee extension L: 12.1 (1.6–30.6). **Goniometry** (21/22): elbow flexion R: 149.2 (130–161); elbow Flexion L: 148.4 (135–165); elbow extension R: −39.8 (−120 to 0); elbow extension L: −34.7 (−114 to 0); knee extension: −15.3 (−95 to 12); knee extension L: −15.6 (−92 to 10). **10 m walk/run** (11/22): 8.6 (4.0–15.8) Year 2 **MFM32** (29/32): total 58.21 (19.8–100). Standing position and transfers (D1): 31.17 (0–100).; axial and proximal motor function (D2): 72.69 (11.1–100); distal (D3): 83.40 (47.6–100). **Myometry** (27/32): elbow flexion R: 5.23 (0–23.8); elbow flexion L: 5.46 (0–19.9); elbow extension R: 3.13 (0–20.1); elbow extension L: 3.42 (0–18.6); knee flexion R: 9.42 (2.3–36.3); knee flexion L: 9.60 (2.3–30.4); knee extension R: 10.80 (2.0–43.6); knee extension L: 10.14 (0–46.9). **Goniometry** (21/22): elbow flexion R: 149.52 (140–160); elbow flexion L: 148.04 (138–160); elbow extension R: −38.09 (−115 to 10); elbow extension L: −37.00 (−120 to 8); knee extension R: −19.77 (−86 to 16); knee extension L: −19.73 (−105 to 10). Stand time: (11/32) **10 m walk/run** (11/32): 7.70 (3.0–14.7); **6MWT** (11/32): 338.27 (144–600); **QUEST** (29/32): 77.68 (28.24-100); **NSAA** (29/32): 7.52 (0–34); **HFMS** (29/32): 22.24 (0–40) The MFM32 total score showed a significant difference between non-ambulatory individuals with a median score of 44.8 (range: 19.8–80.2) and ambulatory individuals with a median score of 82.3 (range: 59.4–100.0). Likewise, the HFMS showed a significant difference with a non-ambulatory median score of 9.5 (range: 0.0–38.0) and an ambulatory median score of 36.0 (range: 22–40).
		COL6-RD	n = 17 (year 2); 9 F/8 M; 12.6 (4.8–22.2) years		
		LAMA2-RD	n = 15 (year 2); 10 F/5 M; 7.9 (5.0–19.3) years		

4. Discussion

The scoping review method enabled an exhaustive literature search, a definition of the current level of knowledge, and the identification of gaps with respect to the current knowledge of motor function in pediatric patients with congenital muscular dystrophies.

This study confirms that data published in the last 20 years on motor outcomes in patients with CMD are limited. Only 16 articles on this topic were found, of which 2 focused on COLVI-RD [8,9], 1 on LMNA-RD [10], 2 on LAMA2-RD [11,12], 2 on FCMD [14,15], 2 on SEPN1-MD [15,16], 1 on FKRP mutations [17], and 6 on different forms of CMD [18–23].

As already reported by Zambon et al., the rarity of these diseases and the great phenotypic heterogeneity make large and comparable studies complex [5]. In this regard, attention is drawn to the pathologies analyzed in the selected articles. Despite the presence of numerous forms of CMD, most of the studies focused on the most represented and well-known forms (COLVI-RD, FCMD, LMNA-RD, LAMA2-RD, SEPN1-MD, FKRP). Only two articles mentioned other forms, although they are grouped in a single analysis category [19,21].

Out of the 16 studies selected, 11 were cross-sectional studies. Only five studies were longitudinal and therefore analyzed the progression of motor function over time [11,15,17,18,22]. The durations of these longitudinal studies varied but were generally short and do not allow broad trend curves to be drawn. Two articles showed results from the same sample of patients to whom different inclusion criteria were applied [18,22]. In fact, 44 of the 48 patients included in the study by Le Goff et al. were included in the study by Jain et al., i.e., all those with a diagnosis of LAMA2-RD or COL6-RD and with the completion of at least two MFM-32 evaluations one year apart. The cross-sectional studies provided information on the motor and functional aspects of the different forms of CMD, describing their common characteristics. However, they do not allow the evolution of the disease over time to be delineated.

The age ranges of the patients varied. Only 8 out of 16 studies showed results from exclusively pediatric patients (0–18 years), although with different age ranges. The other 8 studies included samples of pediatric and adult patients, almost always without distinction in the analysis.

Nevertheless, interesting preliminary data emerged regarding the progression of motor function in patients with LAMA2-RD, COL6-RD, SPN1-RM, and FKRP mutations.

In patients with LAMA2-RD, an annual linear reduction of 6.6° for right elbow flexion and 3.1° for knee flexion was described in the article by Zambon et al. [11], and an annual reduction of 3.21° for left elbow extension in the article by Jain et al. [18]. In the study by Jain et al., a decrement of 2.60 points each year emerged on the MFM32 scale [18]. This decrement was confirmed by the study of Le Goff et al., which showed a greater loss of score in the items related to axial and proximal function (−1.59 points per year) [22].

In patients with COL6-RD, an annual decrease of 4.01 points on the MFM32 scale was described in the study by Jain et al. [18] and confirmed by the study of Le Goff [22]. The domain in which there was the greatest loss of function was that of standing position and transfers (−1.99 points per year) [22].

In patients with SPN1-RM, an annual change of −0.55 points on the HFMS scale was estimated [15].

In patients with FKRP mutations, average annual decreases of 23.41 m at 6MWT, 0.65 points on the MHFMS scale, 1.33 on the PEDI scale, and 0.55 on the PUL scale were reported [17].

Additional larger longitudinal studies differentiated by pediatric age and adulthood with planned follow-up would be recommended to study the course of the diseases over time, better understand the characteristics of the different forms, and trace natural history trajectories. This would also be useful for identifying the outcome measures for trial readiness, as well as implementing innovative care and rehabilitation techniques in clinical practice. This incitement has been partly received by scientific and patient communities. Some disease registries have recently been proposed, such as the Global Registry for COL6-related dystrophies (NCT04020159) or the Global FKRP Registry (NCT04001595), and some natural history studies for clinical trial readiness have been registered such as the study of dystroglycanopathies (NCT00313677) or the study of patients with SEPN1 or LAMA2 mutations (NCT04478981).

Another feature that emerged is the heterogeneity of the assessment tools used in the studies. In 10 out of 16 papers, motor assessment instruments, such as functional scales and time tests, were used (see Table 8). In the other six papers, anamnestic data and information acquired during clinical assessments were analyzed.

Table 8. Assessment tools used in the various selected studies (MFM32: Motor Function Measure; GMFM: Gross Motor Function Measure; HMFS: Hammersmith Motor Function; MHFMS: Modified Hammersmith Motor Function; PEDI: Pediatric Evaluation of Disability Inventory; NSAA: North Star Ambulatory Assessment; TFTs: timed function tests; 6MWT: 6 min walk test; Jebsen: Jebsen–Taylor Hand Function Test; PUL: Performance of Upper Limb; Brooke scale; QUEST: Quality of Upper Extremity Skills Test).

	MFM32	GMFM	HFMS	PEDI	NSAA	TFTs	The NM-Score Classification	6MWT	JEBSEN	PUL	Brooke	QUEST	The Upper Limb Function Measure of Muscular Dystrophy
Natera-de et al., 2021 [9]	x				x	x		x					
Harada et al., 2022 [13]		x	x										x
Silwal et al., 2020 [15]			x										
Gedlinske et al., 2020 [17]			x (modified)	x		x		x		x			
Jain et al., 2019 [18]	x												
Vuillerot et al., 2014 [19]	x						x		x		x		
Bendixen et al., 2017 [20]	x								x			x	
Vuillerot et al., 2014 [21]	x												
Le Goff et al., 2021 [22]	x												
Meilleur et al., 2015 [23]	x		x		x	x		x				x	

Highlighted with 'x' are which rating scales were used in the various studies.

The absence of evaluation instruments specifically validated for these diseases should be underlined. The functional assessment scales reported have been validated for the assessment of patients with neuromuscular disease or other specific neuromuscular diseases or other motor disorders. The scales used were the Motor Function Measure (MFM32), Gross Motor Function Measure (GMFM), Hammersmith Motor Function (HFMS), Modified Hammersmith Motor Function (MHFMS), Pediatric Evaluation of Disability Inventory (PEDI), North Star Ambulatory Assessment (NSAA), timed function tests (TFTs), NM-Score classification, 6 min walk test (6MWT), Jebsen Taylor Hand Function Test (Jebsen), Performance of Upper Limb (PUL), Brooke scale, Quality of Upper Extremity Skills Test (QUEST), and Upper Limb Function Measure of Muscular Dystrophy. The tools cited allow the different motor and functional aspects to be assessed. According to the scale, it is possible to obtain results relating to gross motor functions, fine motor functions, upper limb abilities, fatigability, and speed of execution. In addition, some articles included instruments for measuring joint range and strength.

As can be seen in Table 8, the MFM32 scale was the most used instrument for the assessment of patients with CMD. The subdivision into domains (D1, standing and transfers; D2, axial and proximal motor function; and D3, distal motor function) allows for a comprehensive assessment of motor function [22]. It should be noted that patients with CMD were also included in the validation study of this scale [24].

The study by Meilleur et al. on patients with COL6-RD and LAMA2-RD showed a high association between the MFM32 scale and HFMS. The MFM32 provides more information, so it may be more convenient for the assessment of patients with CMD. However, it should be pointed out that some difficulties related to muscle contractures interfered with some tasks, either due to the inability to reach the initial position or the inability to complete the items [23].

The NSAA scale, which has been validated for patients with Duchenne muscular dystrophy [25] and spinal muscular atrophy type 3 [26], includes the assessment of more complex skills, which is useful for patients with better motor functioning who show ceiling effects on the MFM32 scale. It could, therefore, be a complementary tool [23].

With respect to upper limb function, the QUEST showed associations with the total score from the MFM32 in patients with COL6-RD and LAMA2-RD. However, further investigations are ongoing to understand which of these tools is more appropriate [23].

There were no studies describing the use of functional assessment scales in CMD patients in the neonatal period. However, it would be interesting to identify the most appropriate tool among the Chop Intend scale, the HFMS [27], or the recently proposed module of the Hammersmith Neonatal Neurological Examination (HNEE) for neonates with SMA [28].

Finally, the presence of contractures should be taken into consideration. The most frequently described muscle contractures are at the level of the long flexors of the fingers, elbows, knees, hips, and Achilles tendons. The interference reported in the MFM32 scale scores [23] can be expanded to all functional assessment scales, as the presence of contractures, regardless of the body area, limits functionality. Moreover, in some scales, not reaching the initial position corresponds to not scoring. This could represent a bias for those patients who, although limited by contractures, manage to complete the task anyway by implementing compensation strategies. The contractures may also highlight strength deficits when strength is assessed with the MRC scale or myometry [27].

As already highlighted at the 173rd ENMC international workshop on CMD, there is a need to identify the best motor and functional outcome tools for the assessment of patients with CMD [27].

The scientific community has made efforts in recent years, although studies only concern some of the forms and are still quantitatively limited.

It would be useful to refine research by focusing on additional parameters, such as age, stage of disease, residual motor function, form, and neurophysiological assessments, even if in the related literature, data are limited because of lower levels of compliance in pediatric age. This would make studies conducted in different centres comparable and aggregable.

5. Conclusions

In recent decades, the implementation of standards of care has changed the survival of patients suffering from congenital muscular dystrophies. Moreover, new genetic diagnostic techniques have made it possible to identify an increasing number of different forms.

There is an increasing need for larger longitudinal natural history studies, which are essential to identifying motor outcome measures for trial readiness, as well as implementing innovative treatment and rehabilitation techniques in clinical practice.

As can be seen from this scoping review, the scientific community, especially in recent years, has begun to respond to this challenge. However, it is essential to further expand the knowledge of the forms that have already been studied and proceed with new studies for the forms that are still underinvestigated. Finally, it would be advisable to define the most appropriate evaluation tools so that they can be used in a more homogeneous and, therefore, comparable manner.

Author Contributions: Conceptualization, I.C., T.M., and F.S.R.; methodology, I.C., R.D., C.B., E.P., T.M. and F.S.R.; validation, I.C., R.D., C.B., E.P., F.R, E.R. and F.S.R.; formal analysis, I.C., R.D., C.B. and F.S.R.; investigation, I.C., R.D. and F.S.R.; resources, I.C.; data curation, I.C.; writing—original draft preparation, I.C.; writing—review and editing, I.C., R.D., C.B., E.P., E.R., F.R., T.M. and F.S.R.; visualization, T.M. and F.R.; supervision, T.M. and F.R. All authors have read and agreed to the published version of the manuscript.

Funding: This research received no external funding.

Institutional Review Board Statement: Not applicable.

Informed Consent Statement: Not applicable.

Data Availability Statement: The data presented in this study are openly available in publicly accessible repositories (Pubmed/MEDLINE, Embase).

Conflicts of Interest: The authors declare no conflict of interest.

Appendix A

Pubmed search string

((("Muscular Dystrophies"[Mesh] OR "muscular dystroph*"[tiab]) AND ("congenital"[Subheading] OR congenit*[tiab] OR connatal[tiab] OR CMD[tiab] OR Autosomal[tiab] OR inherit*[tiab] OR inborn[tiab] OR heredit*[tiab])) OR "Muscular Dystrophy, Congenital, due to Partial LAMA2 Deficiency"[Supplementary Concept] OR "Muscular dystrophy congenital, merosin negative"[Supplementary Concept] OR "Bethlem myopathy"[Supplementary Concept] OR "Walker-Warburg Syndrome"[Mesh] OR "Muscular Dystrophy, Emery-Dreifuss"[Mesh] OR alpha-dystroglycanopath*[tiab] OR alpha-DGP[tiab] OR "Walker-Warburg"[tiab] OR "Warburg syndrom*"[tiab] OR "Muscle-Eye-Brain Disease*"[tiab] OR COD-MD[tiab] OR "Emery Dreifuss"[tiab] OR laminopath*[tiab] OR "Bethlem myopath*"[tiab] OR ((Ulrich*[tiab] OR Ullrich*[tiab] OR "laminin alpha 2"[Supplementary Concept] OR laminin-alpha2[tiab] OR LAMA2[tiab] OR "laminin alpha 2"[tiab] OR "Laminin"[Mesh] OR merosin[tiab] "OR Collagen Type VI"[Mesh] OR "collagen VI"[tiab] OR "collagen type VI"[tiab] OR Col6[tiab] OR col6A*[tiab] OR Col-6[tiab] OR col-6A*[tiab] OR "Dystroglycans"[Mesh] OR alpha-dystroglycan*[tiab] OR a-dystroglycan*[tiab] OR FKTN[tiab] OR POMT*[tiab] OR FKRP[tiab] OR POMGNT*[tiab] OR ISPD[tiab] OR B3GNT*[tiab] OR GMPPB[tiab] OR DPM1[tiab] OR DPM2[tiab] OR ALG13[tiab] OR B3GALNT*[tiab] OR RXYLT*[tiab] OR "Lamin Type A"[Mesh] OR "lamin A-C"[tiab] OR "lamin A/C"[tiab] OR LMNA[tiab]) AND dystroph*[tiab])) AND ("Motor Skills"[Mesh] OR "Range of Motion, Articular"[Mesh] OR "Movement"[Mesh] OR motion[tiab] OR motor[tiab] OR movement*[tiab] OR locomot*[tiab] OR mobility[tiab] OR ambulat*[tiab] OR walking*[tiab] OR "Scoliosis"[Mesh] OR scolios*[tiab] OR "Natural History"[Mesh] OR "natural history"[tiab]).

Embase search string

('motor performance'/exp OR 'motor skills in infancy and childhood'/exp OR 'movement (physiology)'/exp OR 'scoliosis'/exp OR 'history'/exp OR motion:ab,ti OR motor:ab,ti OR movement*:ab,ti OR locomot*:ab,ti OR mobility:ab,ti OR ambulat*:ab,ti OR walking*:ab,ti OR 'natural hystory':ab,ti OR scolios*:ab,ti) AND (('muscular dystrophy'/exp AND (congenit*:ab,ti OR connatal:ab,ti OR cmd:ab,ti OR autosomal:ab,ti OR inherit*:ab,ti OR inborn*:ab,ti OR heredit*:ab,ti) OR 'congenital muscular dystrophy type 1a'/exp OR 'merosin deficient congenital muscular dystrophy'/exp OR 'merosin'/exp OR 'bethlem myopathy'/exp OR 'walker warburg syndrome'/exp OR 'emery dreifuss muscular dystrophy'/exp OR 'alpha dystroglycanopathy'/exp OR 'muscle eye brain disease'/exp OR 'laminopathy'/exp OR 'ullrich congenital muscular dystrophy'/exp OR 'laminin'/exp OR 'laminin alpha2'/exp OR 'collagen type 6'/exp OR 'dystroglycan'/exp OR 'alpha dystroglycan'/exp OR 'lamin a'/exp OR 'alpha dystroglycanopath*':ab,ti OR 'alpha dgp':ab,ti OR 'walker-warburg':ab,ti OR 'warburg syndrom*':ab,ti OR 'muscle-eye-brain disease*':ab,ti OR 'cod md':ab,ti OR 'emery dreifuss':ab,ti OR laminopath*:ab,ti OR 'bethlem myopath*':ab,ti

OR ulrich*:ab,ti OR ullrich*:ab OR 'laminin alpha2':ab,ti OR lama2:ab,ti OR 'laminin alpha 2':ab,ti OR merosin:ab,ti OR 'collagen vi':ab,ti OR 'collagen type vi':ab,ti OR col6:ab,ti OR col6a*:ab,ti OR 'col 6':ab,ti OR 'col 6a*':ab,ti OR 'alpha dystroglycan*':ab,ti OR 'a dystroglycan*':ab,ti OR fktn:ab,ti OR pomt*:ab,ti OR fkrp:ab,ti OR pomgnt*:ab,ti OR ispd:ab,ti OR b3gnt*:ab,ti OR gmppb:ab,ti OR dpm1:ab,ti OR dpm2:ab,ti OR alg13:ab,ti OR b3galnt*:ab,ti OR rxylt*:ab,ti OR 'lamin a-c':ab,ti OR 'lamin a/c':ab,ti OR lmna:ab,ti) AND dystroph*:ab,ti).

References

1. Pasrija, D.; Tadi, P. Congenital Muscular Dystrophy. In *Treasure Island*; StatPearls Publishing: Tampa, FL, USA, 2022. Available online: http://www.ncbi.nlm.nih.gov/books/NBK558956/ (accessed on 20 October 2022).
2. Bertini, E.; D'Amico, A.; Gualandi, F.; Petrini, S. Congenital Muscular Dystrophies: A Brief Review. *Semin. Pediatr. Neurol.* **2011**, *18*, 277–288. [CrossRef] [PubMed]
3. Butterfield, R.J. Congenital Muscular Dystrophy and Congenital Myopathy. *Contin. Lifelong Learn. Neurol.* **2019**, *25*, 1640–1661. [CrossRef]
4. Graziano, A.; Bianco, F.; D'Amico, A.; Moroni, I.; Messina, S.; Bruno, C.; Pegoraro, E.; Mora, M.; Astrea, G.; Magri, F.; et al. Prevalence of congenital muscular dystrophy in Italy: A population study. *Neurology* **2015**, *84*, 904–911. [CrossRef]
5. Zambon, A.A.; Muntoni, F. Congenital muscular dystrophies: What is new? *Neuromuscul. Disord.* **2021**, *31*, 931–942. [CrossRef] [PubMed]
6. Tricco, A.C.; Lillie, E.; Zarin, W.; O'Brien, K.K.; Colquhoun, H.; Levac, D.; Moher, D.; Peters, M.D.J.; Horsley, T.; Weeks, L.; et al. PRISMA Extension for Scoping Reviews (PRISMA-ScR): Checklist and Explanation. *Ann. Intern. Med.* **2018**, *169*, 467–473. [CrossRef] [PubMed]
7. The Joanna Briggs Institute. *Joanna Briggs Institute Reviewers' Manual: 2015 Edition/Supplement Adelaide*; The Joanna Briggs Institute: Adelaide, Australia, 2015.
8. Yonekawa, T.; Komaki, H.; Okada, M.; Hayashi, Y.K.; Nonaka, I.; Sugai, K.; Sasaki, M.; Nishino, I. Rapidly progressive scoliosis and respiratory deterioration in Ullrich congenital muscular dystrophy. *J. Neurol. Neurosurg. Psychiatry* **2013**, *84*, 982–988. [CrossRef]
9. Benito, D.N.-D.; Foley, A.R.; Domínguez-González, C.; Ortez, C.; Jain, M.; Mebrahtu, A.; Donkervoort, S.; Hu, Y.; Fink, M.; Yun, P.; et al. Association of Initial Maximal Motor Ability With Long-term Functional Outcome in Patients With COL6-Related Dystrophies. *Neurology* **2021**, *96*, e1413–e1424. [CrossRef]
10. Fan, Y.; Tan, D.; Song, D.; Zhang, X.; Chang, X.; Wang, Z.; Zhang, C.; Chan, S.H.-S.; Wu, Q.; Wu, L.; et al. Clinical spectrum and genetic variations of *LMNA*-related muscular dystrophies in a large cohort of Chinese patients. *J. Med. Genet.* **2020**, *58*, 326–333. [CrossRef]
11. Zambon, A.A.; Ridout, D.; Main, M.; Mein, R.; Phadke, R.; Muntoni, F.; Sarkozy, A. LAMA2-related muscular dystrophy: Natural history of a large pediatric cohort. *Ann. Clin. Transl. Neurol.* **2020**, *7*, 1870–1882. [CrossRef]
12. Tan, D.; Ge, L.; Fan, Y.; Chang, X.; Wang, S.; Wei, C.; Ding, J.; Liu, A.; Wang, S.; Li, X.; et al. Natural history and genetic study of LAMA2-related muscular dystrophy in a large Chinese cohort. *Orphanet J. Rare Dis.* **2021**, *16*, 319. [CrossRef]
13. Harada, R.; Taniguchi-Ikeda, M.; Nagasaka, M.; Nishii, T.; Inui, A.; Yamamoto, T.; Morioka, I.; Kuroda, R.; Iijima, K.; Nozu, K.; et al. Assessment of the upper limb muscles in patients with Fukuyama muscular dystrophy: Noninvasive assessment using visual ultrasound muscle analysis and shear wave elastography. *Neuromuscul. Disord.* **2022**, *32*, 754–762. [CrossRef]
14. Ishigaki, K.; Ihara, C.; Nakamura, H.; Mori-Yoshimura, M.; Maruo, K.; Taniguchi-Ikeda, M.; Kimura, E.; Murakami, T.; Sato, T.; Toda, T.; et al. National registry of patients with Fukuyama congenital muscular dystrophy in Japan. *Neuromuscul. Disord.* **2018**, *28*, 885–893. [CrossRef] [PubMed]
15. Silwal, A.; Sarkozy, A.; Scoto, M.; Ridout, D.; Schmidt, A.; Laverty, A.; Henriques, M.; D'Argenzio, L.; Main, M.; Mein, R.; et al. Selenoprotein N-related myopathy: A retrospective natural history study to guide clinical trials. *Ann. Clin. Transl. Neurol.* **2020**, *7*, 2288–2296. [CrossRef] [PubMed]
16. Villar-Quiles, R.N.; von der Hagen, M.; Métay, C.; Gonzalez, V.; Donkervoort, S.; Bertini, E.; Castiglioni, C.; Chaigne, D.; Colomer, J.; Cuadrado, M.L.; et al. The clinical, histologic, and genotypic spectrum of *SEPN1*-related myopathy. *Neurology* **2020**, *95*, e1512–e1527. [CrossRef]
17. Gedlinske, A.M.; Stephan, C.M.; Mockler, S.R.; Laubscher, K.M.; Laubenthal, K.S.; Crockett, C.D.; Zimmerman, M.B.; Mathews, K.D. Motor outcome measures in patients with *FKRP* mutations. *Neurology* **2020**, *95*, e2131–e2139. [CrossRef] [PubMed]
18. Jain, M.S.; Meilleur, K.; Kim, E.; Norato, G.; Waite, M.; Nelson, L.; McGuire, M.; Duong, T.; Keller, K.; Lott, D.J.; et al. Longitudinal changes in clinical outcome measures in COL6-related dystrophies and LAMA2-related dystrophies. *Neurology* **2019**, *93*, e1932–e1943. [CrossRef] [PubMed]
19. Vuillerot, C.; Meilleur, K.G.; Jain, M.; Waite, M.; Wu, T.; Linton, M.; Datsgir, J.; Donkervoort, S.; Leach, M.E.; Rutkowski, A.; et al. English Cross-Cultural Translation and Validation of the Neuromuscular Score: A System for Motor Function Classification in Patients With Neuromuscular Diseases. *Arch. Phys. Med. Rehabilitation* **2014**, *95*, 2064–2070.e1. [CrossRef]

20. Bendixen, R.M.; Butrum, J.; Jain, M.S.; Parks, R.; Hodsdon, B.; Nichols, C.; Hsia, M.; Nelson, L.; Keller, K.C.; McGuire, M.; et al. Upper extremity outcome measures for collagen VI-related myopathy and LAMA2-related muscular dystrophy. *Neuromuscul. Disord.* **2016**, *27*, 278–285. [CrossRef]
21. Vuillerot, C.; Rippert, P.; Kinet, V.; Renders, A.; Jain, M.; Waite, M.; Glanzman, A.M.; Girardot, F.; Hamroun, D.; Iwaz, J.; et al. Rasch Analysis of the Motor Function Measure in Patients With Congenital Muscle Dystrophy and Congenital Myopathy. *Arch. Phys. Med. Rehabilitation* **2014**, *95*, 2086–2095. [CrossRef]
22. Le Goff, L.; Meilleur, K.G.; Norato, G.; Rippert, P.; Jain, M.; Fink, M.; Foley, A.R.; Waite, M.; Donkervoort, S.; Bönnemann, C.G.; et al. Responsiveness and Minimal Clinically Important Difference of the Motor Function Measure in Collagen VI-Related Dystrophies and Laminin Alpha2-Related Muscular Dystrophy. *Arch. Phys. Med. Rehabilitation* **2020**, *102*, 604–610. [CrossRef]
23. Meilleur, K.G.; Jain, M.S.; Hynan, L.S.; Shieh, C.-Y.; Kim, E.; Waite, M.; McGuire, M.; Fiorini, C.; Glanzman, A.M.; Main, M.; et al. Results of a two-year pilot study of clinical outcome measures in collagen VI- and laminin alpha2-related congenital muscular dystrophies. *Neuromuscul. Disord.* **2014**, *25*, 43–54. [CrossRef] [PubMed]
24. Bérard, C.; Payan, C.; Hodgkinson, I.; Fermanian, J.; The MFM Collaborative Study Group A Motor Function Measure Scale for Neuromuscular Diseases. Construction and validation study. *Neuromuscul. Disord.* **2005**, *15*, 463–470. [CrossRef]
25. Scott, E.; Eagle, M.; Mayhew, A.; Freeman, J.; Main, M.; Sheehan, J.; Manzur, A.; Muntoni, F.; Disease, T.N.S.C.N.F.P.N. Development of a Functional Assessment Scale for Ambulatory Boys with Duchenne Muscular Dystrophy. *Physiother. Res. Int.* **2011**, *17*, 101–109. [CrossRef] [PubMed]
26. Tiziano, F.D.; Lomastro, R.; Di Pietro, L.; Pasanisi, M.B.; Fiori, S.; Angelozzi, C.; Abiusi, E.; Angelini, C.; Sorarù, G.; Gaiani, A.; et al. Clinical and molecular cross-sectional study of a cohort of adult type III spinal muscular atrophy patients: Clues from a biomarker study. *Eur. J. Hum. Genet.* **2012**, *21*, 630–636. [CrossRef] [PubMed]
27. Bönnemann, C.; Rutkowski, A.; Mercuri, E.; Muntoni, F. 173rd ENMC international workshop: Congenital muscular dystrophy outcome measures 5–7 March 2010, Naarden, The Netherlands. *Neuromuscul. Disord.* **2011**, *21*, 513–522. [CrossRef]
28. Pane, M.; Donati, M.A.; Cutrona, C.; De Sanctis, R.; Pirinu, M.; Coratti, G.; Ricci, M.; Palermo, C.; Berti, B.; Leone, D.; et al. Neurological assessment of newborns with spinal muscular atrophy identified through neonatal screening. *Eur. J. Pediatr.* **2022**, *181*, 2821–2829. [CrossRef]

applied sciences

MDPI

Article

Oral Immune-Related Adverse Events Associated with PD-1 Inhibitor Treatment: A Case Series

Daniele Pergolini [1], Andrea Botticelli [2], Roberta Fascetti [1], Federica Rocchetti [1], Alessio Cirillo [2], Gianluca Tenore [1], Gaspare Palaia [1,*], Antonella Polimeni [1] and Umberto Romeo [1]

[1] Department of Oral and Maxillofacial Sciences, Sapienza University of Rome, Via Caserta 6, 00161 Rome, Italy
[2] Department of Radiological, Oncological and Anatomical-Pathological Sciences, Sapienza University of Rome, Viale Del Policlinico 155, 00161 Rome, Italy
* Correspondence: gaspare.palaia@uniroma1.it; Tel.: +39-452-2515

Abstract: Introduction: Immune Checkpoint Inhibitor (ICI) drugs have led to a revolution in the treatment of different forms of cancer, shifting the target of action from cancer cells to the patient's immune system, enhancing their responses against the tumor itself. On the other hand, this mechanism can lead to responses against oneself, with the appearance of immune-related adverse events. The aim of the present study was to examine the immune-related adverse events (irAEs) affecting the mucous membranes of the oral cavity and the possible correlation between these and skin toxicities, which are reported in the literature as the most common adverse events. Materials and methods: Thirteen patients treated with anti-Programmed Death (PD-1) drugs (pembrolizumab, nivolumab, and cemiplimab) were selected. The data collected include the general history of the patient and the type of anticancer treatment. The sample was then analyzed by recording the alterations found on the mucous membranes of the oral cavity and on the skin. Finally, the average time that elapsed between the start of immunotherapy and the onset of lesions was analyzed. Results: Patients often had multiple lesions at the same time. Hyperkeratosis was found in three patients, candidiasis (pseudomembranous and median rhomboid glossitis) in two patients, epithelial atrophy in four patients, and ulcerative areas in two patients. One patient reported xerostomia with dysphagia. The anatomical areas most involved were the dorsal tongue and palate. Skin irAEs included skin rash erythema ($n = 7$) with diffuse redness, the presence of small bubbles with a crusty outcome, and dryness of the skin in the affected areas. Discussion: In the literature, there are few studies that analyze how irAEs affect the mucous membranes of the oral cavity in patients treated with ICI drugs. The most frequently described lesions are lichenoid reactions and xerostomia. Moreover, the development of mucositis, generally of low grade, has been reported. The present study has confirmed the data from the literature and, in addition, reports two cases of candidiasis, an adverse event that has never been shown in the literature. Conclusions: irAEs have the potential to affect any organ. The only way to avoid the occurrence of serious events that is currently available is early interception, which is only possible through the knowledge of these manifestations. It is therefore considered necessary to deepen our knowledge of oral irAEs and their correlation with dermatological toxicities, allowing for a multidisciplinary classification of the patient and a timely diagnosis of any adverse event and avoiding progression to more advanced stages, which could lead to the temporary or permanent suspension of anticancer drugs.

Keywords: cancer therapy; immunotherapy; oral adverse events; oral disease; oral lesions

Citation: Pergolini, D.; Botticelli, A.; Fascetti, R.; Rocchetti, F.; Cirillo, A.; Tenore, G.; Palaia, G.; Polimeni, A.; Romeo, U. Oral Immune-Related Adverse Events Associated with PD-1 Inhibitor Treatment: A Case Series. *Appl. Sci.* **2022**, *12*, 12994. https://doi.org/10.3390/app122412994

Academic Editors: Tommaso Lombardi and Andrea Scribante

Received: 24 October 2022
Accepted: 14 December 2022
Published: 18 December 2022

1. Introduction

Immune Checkpoint Inhibitor (ICI) drugs are a new category of anticancer drugs able to act on the patient's immune system, enhancing its response against cancer cells and restoring the state of immunosurveillance [1,2]. Responses obtained through this approach have led to objective results, such as reduction of tumor mass, and to an increase in survival compared to traditional therapies [1].

Another characteristic that distinguishes immunotherapy is the ability to establish an immunological memory, allowing it to have lasting responses over time, improving the progression-free survival of the disease and overall survival [3].

Nowadays, the most popular drugs in cancer immunotherapy are monoclonal antibodies, which act at two immune checkpoints CTLA-4 (Cytotoxic T-Lymphocyte-Associated Antigen 4) and PD-1/PDL-1 (Programmed Cell Death Protein-1), receptors with immunosuppressive functions. The receptor–antibody binding inhibits the immunosuppressive signal, allowing T-lymphocytes to continue to proliferate and play a defensive role in the tumor microenvironment [4,5]. Although antitumor cytotoxic activity is therefore favored, this mechanism of action may lead to the onset of immune-related adverse events (irAEs) [6]. The organs that are the most involved are the colon, liver, lungs, pituitary gland, thyroid, and skin, although less common adverse events have been described affecting the heart, nervous system, and other organs [7].

There are many studies in the literature that analyze irAEs, but few are related to the oral mucous membranes. In this study, oral manifestations were examined during immunotherapy with ICI drugs, with anti-PD-1 drugs in particular. The aim of the study was to evaluate the type of lesions, the timing with which they occur during treatment, and the response to treatment. In addition, since the reported cutaneous irAEs had the highest incidence rate, a possible correlation between skin changes and intraoral manifestations was examined.

2. Materials and Methods

Thirteen patients treated with anti-PD-1 drugs (pembrolizumab, nivolumab, and cemiplimab) were selected. Two patients were excluded from the study due to discontinuation of immunotherapy following the first administration. Data were collected on the general history of each patient: age, sex, remote pathological history with any drugs taken, and tumor diagnosis. In addition, information related to immunotherapy treatment was collected: type of drug, dosage of the drug, date of first infusion, and frequency of infusions. It was reported whether the patient had undergone combined treatment with chemotherapy or radiotherapy. The sample was then analyzed by recording the alterations found. Finally, the median time between the start of immunotherapy and the onset of lesions on the oral mucous membranes and skin were analyzed in relation to the number of infusions.

Patients were evaluated after the first administration (average duration: 19.3 days) of the immunosuppressant drug. Follow-up visits were carried out following subsequent infusions, up to the 10th infusion. On each visit, any changes to the oral mucous membranes or skin were reported (Table 1).

Table 1. Patient data.

Patient	County′	Sex	Cancer Diagnosis	Therapy Anti PD-1	Dosage	Smoke	Radiotherapy	Chemotherapy	Pathological History
1	64	F	K urothelial	pembrolizumab	200 mg flat dose q21	Yes	No	No	Systemic lupus erythematosus
2	67	M	K parotid	pembrolizumab	200 mg flat dose q21	Former smoker	10 sessions	No	Hypertension
3	76	M	Pulmonary ADK	pembrolizumab	200 mg flat dose q21	No	No	No	Renal failure, aortic stenosis
4	68	M	K squamous skin	cemiplimab	350 mg dropped dose q 21	No	2 sessions	No	Rheumatoid arthritis
5	36	M	Melanoma	nivolumab	240 mg q 14	No	No	No	Nothing to detect
6	81	M	K squamous cell oral cavity	pembrolizumab	200 mg flat dose q21	Former smoker	No	No	Hypertension; diabetes II; dyslipidemia
7	68	F	K squamous cell oral cavity	pembrolizumab	200 mg flat dose q21	No	No	No	Diabetes II; hypertension; chronic HBV; diverticular pathology
8	70	M	K urothelial	pembrolizumab	200 mg flat dose q21	Yes	No	Cisplatin-gemcitabine	Nothing to detect
9	63	M	Lung adenocarcinoma	pembrolizumab	200 mg flat dose q21	No	No	No	Atrial fibrillation
10	69	M	Pulmonary adenocarcinoma	nivolumab	240 mg q 14	Former smoker	No	Cisplatin alimta	Hypertension; dyslipidemia; partial thyroidectomy; coronary stent
11	66	F	Lung adenocarcinoma	pembrolizumab	200 mg flat dose q21	No	No	Cisplatin pemetrexed	Nothing to detect
12	75	M	Pulmonary adenocarcinoma	pembrolizumab	200 mg flat dose q21	No	No	No	Nothing to detect
13	71	M	Pulmonary adenocarcinoma	pembrolizumab	200 mg flat dose q21	No	10 sessions	No	Hypertension

3. Results

The mean age of patients was 67 years (range: 36–81). Cancer diagnoses included lung adenocarcinoma (n = 6), urothelial cancer (n = 2), oral squamous cell carcinoma (n = 2), parotid cancer (n = 1), squamous skin cancer (n = 1), and melanoma (n = 1). Ten patients were treated with pembrolizumab at a dosage of 200 mg every three weeks; two patients received treatment with nivolumab at a dosage of 240 mg every two weeks; and only one patient was treated with cemiplimab at a dosage of 350 mg every three weeks. In addition, three patients received adjuvant treatment with radiotherapy, and the other three patients received chemotherapy.

3.1. Oral irEAs

Oral immune-related adverse events are reported in Table 2.

Table 2. Oral immune-related adverse events.

Patient	Clinical Description	Type of Lesion	Anatomical Site	Therapy	Anti-PD-1 Dose for Oral irAEs
1	Nothing to detect	Nothing to detect	Nothing to detect	Nothing to detect	
2	Erythematous area	Epithelial atrophy without dysplasia	Palate	Aminogam® mouthwash	1st dose
3	Nothing to detect	Nothing to detect	Nothing to detect	Nothing to detect	
4	Removable whitish plates Erythematous area Xerostomia	Candidiasis (pseudomembranous candidiasis, median rhomboid glossitis)	Dorsal tongue Oral mucous membranes	Nystatin Mucosamin® mouthwash	3rd dose 9th dose
5	Nothing to detect	Nothing to detect	Nothing to detect	Nothing to detect	
6	Non-removable whitish lesions	Hyperkeratosis	Palate Cheek mucosa	Aminogam® mouthwash	4th dose
7	Non-removable white lesions	Hyperkeratosis	Dorsal tongue Palate	Aminogam® mouthwash	3rd dose
8	Nothing to detect	Nothing to detect	Nothing to detect	Nothing to detect	Nothing to detect
9	Nothing to detect	Nothing to detect	Nothing to detect	Nothing to detect	
10	Erythematous and ulcerative areas	Epithelial atrophy without dysplasia Ulcer	Palate		2nd dose
11	Nothing to detect	Nothing to detect	Nothing to detect	Nothing to detect	Nothing to detect
12	Non-removable whitish lesions	Epithelial atrophy and hyperkeratosis without dysplasia	Cheek Dorsal tongue	Aminogam® mouthwash	4th dose
13	Erythematous area Ulcer Removable whitish plates	Epithelial atrophy without dysplasia Ulcer Candidiasis	Palate Tuberosity Dorsal tongue Alveolar process	Nystatin	2nd dose 4th dose

Patients often had multiple lesions at the same time. Hyperkeratosis was found in three patients, candidiasis (pseudomembranous and median rhomboid glossitis) in two patients, epithelial atrophy in four patients, and ulcerative areas in two patients. One patient reported xerostomia with dysphagia. The anatomical areas most involved were the dorsal tongue and palate.

The sites affected by hyperkeratosis included the palate, cheek mucosa, and dorsal tongue (Figure 1). Candidiasis (Figures 2 and 3) presented itself with a pseudomembranous type (Figure 2) or as a median rhomboid glossitis (Figure 3). Erythematous areas (Figure 4), with epithelial atrophy, as well as ulcerative lesions (Figures 5 and 6), were found mainly on the palate.

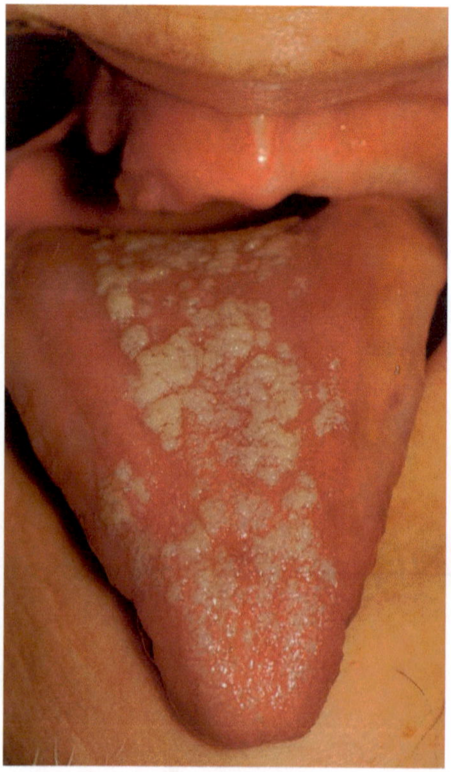

Figure 1. Patchy, erythematous depapillation of the dorsal tongue with regions of elongated filliform papillae. Histological examination showed a picture of atrophy as well as a hyperkeratosis without dysplasia (Patient no. 12, Table 2).

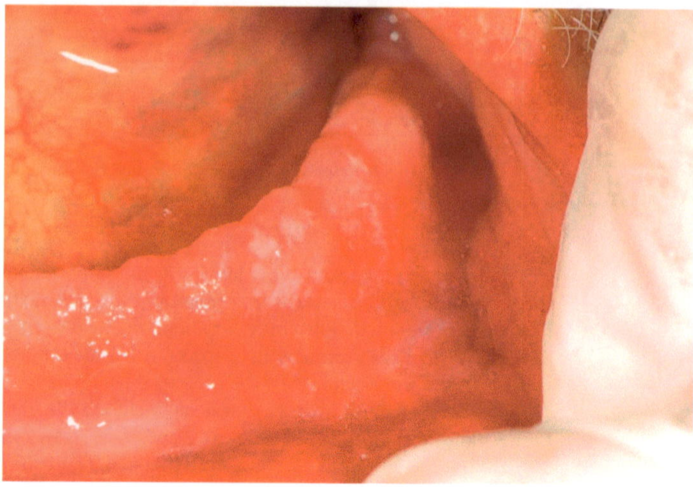

Figure 2. Removable whitish plaques associated with pseudomembranous candidiasis (Patient no. 4, Table 2).

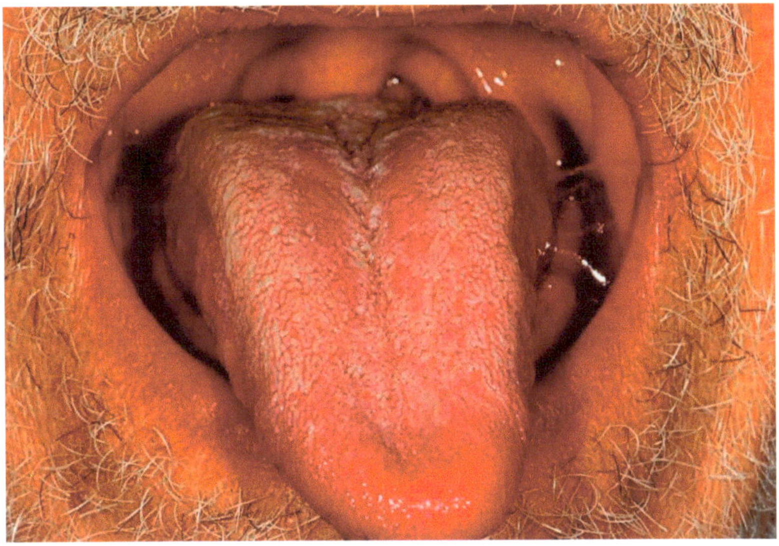

Figure 3. Atrophic area of dorsal tongue associated with median rhomboid glossitis (Patient no. 4, Table 2).

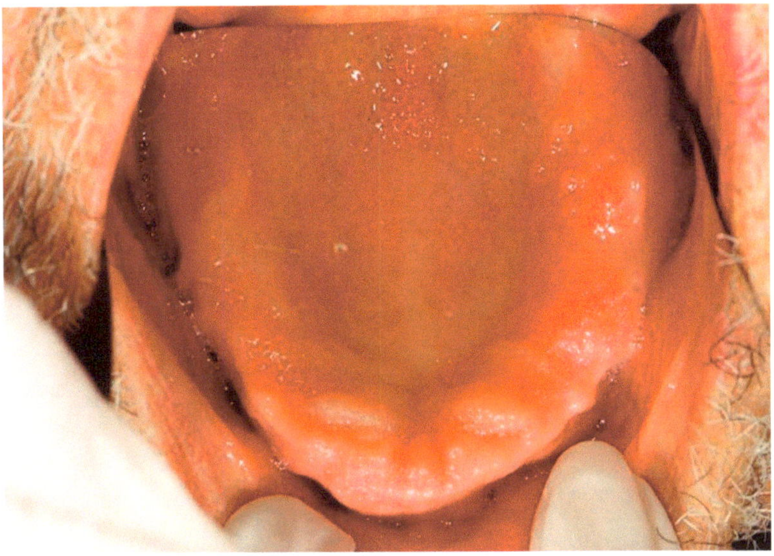

Figure 4. Mucosal erythema, palate (Patient no. 2, Table 2).

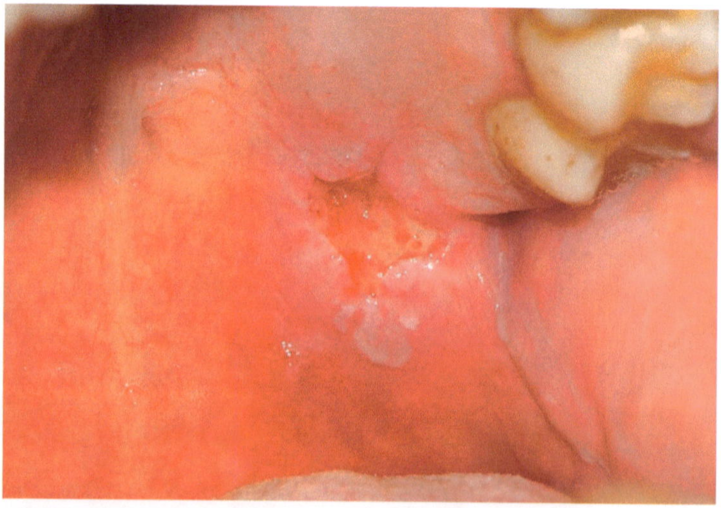

Figure 5. Ulcerative lesion of the hard palate (Patient no. 13, Table 2).

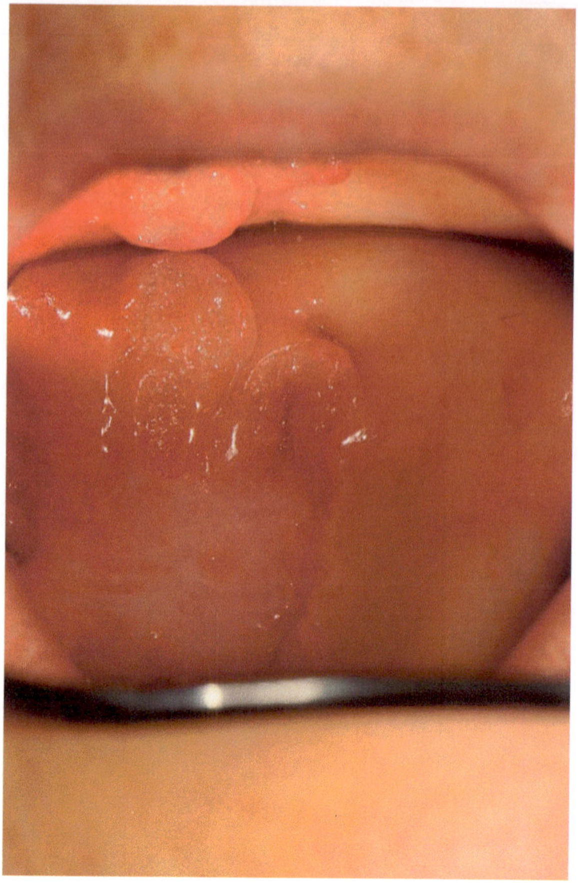

Figure 6. Ulcerative area of the palate (Patient no. 10, Table 2).

Generally, oral alterations were reported at a minimum after the second infusion; in only one patient was an atrophic area in the palate reported after the first dose of pembrolizumab. The event with the earliest onset was found to be mucosal erythema; hyperkeratotic lesions developed between the third/fourth infusion of anti-PD-1 inhibitor therapy. The same timing was reported for the development of candidiasis of the oral mucous membranes, while xerostomia was found later, following the ninth infusion.

Oral candidiasis was diagnosed after an oral swab and a culture test in order to report the amount of the infectious load.

All the nonremovable whitish and reddish lesions were subjected to biopsy and histological examination.

The administration of steroids was generally avoided due to the immunological status of the patients. Ulcerative and erythematous lesions showed complete recovery 10–15 days after the topical administration of hyaluronic acid. Candidiasis was treated with the administration of antifungal drugs (nystatin).

Finally, six patients did not show significant oral alterations.

3.2. Cutaneous irAEs

In the skin, the only adverse event reported was skin rash erythema (Figures 7 and 8) found in seven patients. The clinical manifestations included widespread redness, small blisters with the development of crusts as a result of maceration (Figures 9 and 10), and dryness of the skin in the affected areas. Patients also complained of moderate-to-intense itching. The most frequently affected areas are the abdomen, upper and lower limbs, back, face, and chest. All patients, following dermatological consultation, were treated with topical steroids and reported clinical and symptomatic improvement.

In five patients, skin erythema developed following the third infusion; in only two cases, a more rapid onset was recorded, following the first and second infusions (Table 3).

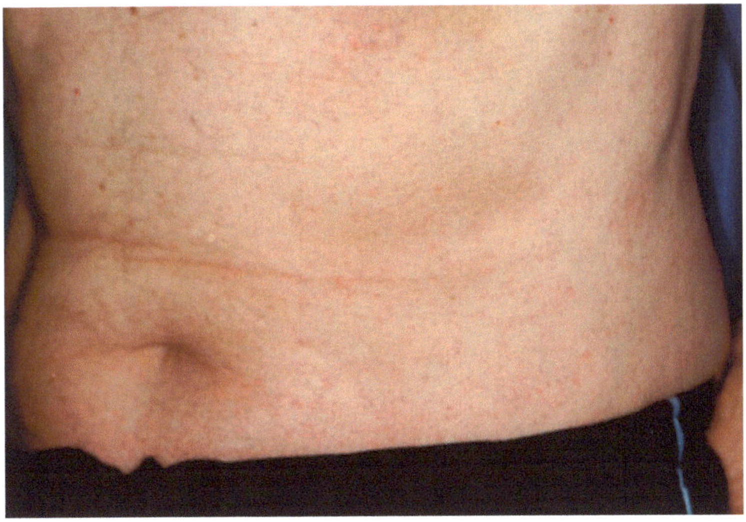

Figure 7. Skin rash erythema (Patient no. 7, Table 3).

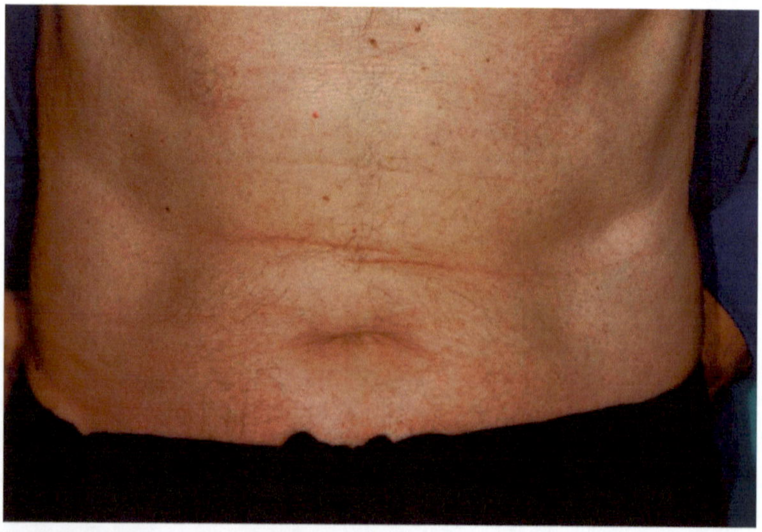

Figure 8. Skin rash erythema (Patient no. 7, Table 3).

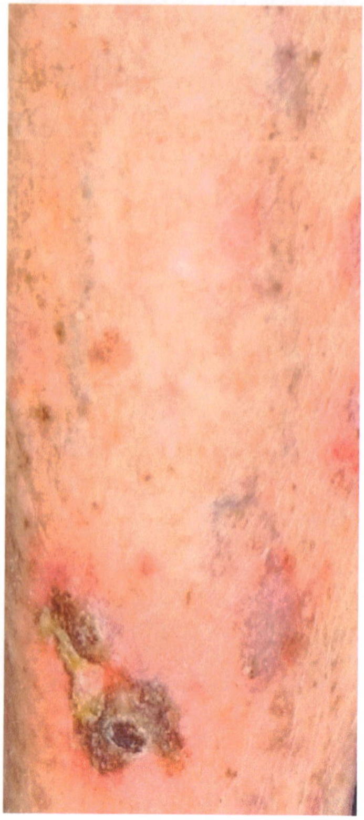

Figure 9. Leg erythema (Patient no. 6, Table 3).

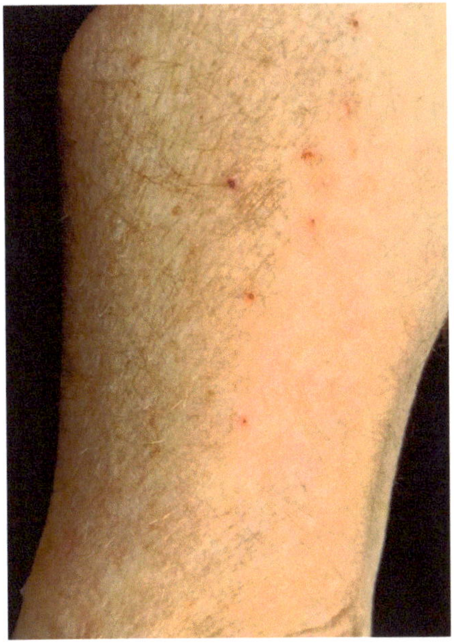

Figure 10. Arm erythema (Patient no. 10, Table 3).

Table 3. Skin immune-related adverse events.

Patient	Lesion	Body Area	Treatment	Dermatological Anti-PD-1 Dose
1	Nothing to detect	Nothing to detect	Nothing to detect	Nothing to detect
2	Nothing to detect	Nothing to detect	Nothing to detect	Nothing to detect
3	Nothing to detect	Nothing to detect	Nothing to detect	Nothing to detect
4	ERYTHEMA	BACK	TOPICAL STEROID	3rd dose
5	Nothing to detect	Nothing to detect	Nothing to detect	Nothing to detect
6	ERYTHEMA	LEGS	TOPICAL STEROID	3rd dose
7	ERYTHEMA	ABDOMEN	TOPICAL STEROID	3rd dose
8	Nothing to detect	Nothing to detect	Nothing to detect	Nothing to detect
9	Nothing to detect	Nothing to detect	Nothing to detect	Nothing to detect
10	ERYTHEMA	ARMS, LEGS, ABDOMEN	TOPICAL STEROID	3rd dose
11	ERYTHEMA	FACE, CHEST	TOPICAL STEROID	3rd dose
12	ERYTHEMA	ABDOMEN	TOPICAL STEROID	1st dose
13	ERYTHEMA	ABDOMEN	TOPICAL STEROID	2nd dose

4. Discussion

The multiple toxicities reported following treatment with ICI drugs require a multidisciplinary approach. Early interception, with resolution or improvement of the lesion, could avoid the interruption of the immunotherapy and the management of severe sequelae.

In terms of overall toxicity, PD-1/PDL-1 inhibitor therapy has a better safety profile than drugs directed against the CTLA-4 receptor. The difference in behavior would seem to depend on the different localization of the two receptors: while CTLA-4 is present in T cells and regulatory T cells (T-REG) and its activation inhibits the immune response at an early stage, the PD-1 receptor is expressed mainly in peripheral tissues and in the tumor environment. The PD-1 inhibitor appears to target T cells with greater specificity [8]. For these reasons, CTLA-4 inhibitors lead to a broader immune response, with a greater occurrence of adverse events than PD-1/PDL-1 inhibitors. Immune-related adverse events in the skin and oral mucous membranes seem to have a different trend: these manifestations are more associated with treatment with anti-PD-1 drugs than with anti-CTLA-4 drugs [9]. From the analysis of the literature and our clinical experience, the correlation between therapy with ICI drugs and the onset of irAEs appears certain. The duration of administration may influence the onset of irAEs, since we have seen that episodes often occur after the first administration of the drug. It should also be emphasized that the aforementioned therapy can, sometimes, be combined with other types of treatments (e.g., chemotherapy), which often lead to a greater appearance of adverse reactions in many areas. It should be expected that the adverse reactions associated with the combined use of ICI drugs and chemotherapy may be more difficult to cure than with just one of these causal factors.

The present study, in accordance with the scientific literature, reported lichenoid reactions among the most frequent oral alterations associated with PD-1-born therapy. Clinically, they appear as whitish and reticulated plaques (Wickham's striae), located mainly at the level of the cheek mucosa, lingual dorsum, and palate. They are generally asymptomatic and occur unilaterally, unlike oral lichen planus, where the distribution is characteristically bilateral. The onset is between the third and fourth doses of the drug, while later onset is described in the literature [10]. In addition, several studies have described the presence of erosive and atrophic elements in the context of whitish plaques that can lead to painful symptoms [11].

In three patients, erythematous areas of the cheek mucosa were found to appear at an early stage of treatment, between the first and second infusion of the drug. In the literature, such manifestations have been described as mucositis and concomitant appearance of ulcerative lesions [12]. They generally present low severity [1,2], while cases of severe immune mucositis associated with esophagitis are rare [13]. Among the examined patients, two reported the appearance of ulcerative lesions in the context of erythematous areas on the palate. Patients reported no painful symptoms as a result of these alterations. Following the appearance of the ulcerative lesion, only one patient complained of mild burning.

While no cases of candidiasis associated with patients receiving anti-PD-1 drugs are reported in the literature, candida infections were recorded in two cases, between the third and fourth infusions of the drug. In the first case, the fungal infection was not associated with other alterations in the oral mucous membranes, while in the second case, it arose in an inflammatory context. In a patient with skin cancer, the candidiasis and xerostomia could also be due to the radiation regimen.

Xerostomia is reported in the literature as the second most frequent oral adverse event following treatment with ICI, described as having an incidence of 6–7.2% [14]. The present study found only one case of xerostomia associated with difficulty swallowing. Often, the dryness of the mucous membranes also causes dysgeusia. Other reported symptoms include thick or sticky saliva, dry throat with hoarseness, and sensitivity to spicy and acidic foods. Cases of dry mouth with a high degree of severity have also been reported, which required temporary or permanent discontinuation of the ICI drug [15].

The most frequent oral adverse events, following treatment with anti-PD-1 drugs, therefore include mucositis, xerostomia, dysgeusia, and lichenoid reactions. Only one study described a case of pemphigoid of mucous membranes following therapy with pembrolizumab (PD-1) [16].

In the present study, six of seven patients with cutaneous irAEs also developed oral alterations. Only in two patients was this correlation not found: in one case, the presence of erythema on the face and chest was found, without any oral manifestation, while one patient reported inflammation of the palatal mucosa in the absence of skin changes. Moreover, adverse reactions were found in four of the six patients with lung cancer and treatment with ICI drugs. It could be analyzed, with a larger number of patients, whether the type of cancer being treated could be related not only to the onset but also to the type and location of the lesions that have arisen.

All patients who underwent radiotherapy developed oral lesions. Skin lesions were found in two of the three patients who received chemotherapy. The lesions found in the latter patients were always of the erythematous type, while the lesions in patients with radiotherapy were of both the erythematous and hyperkeratotic type.

ICI drugs have enabled a revolution in the treatment of cancer diseases, particularly for advanced cancers that are not responsive to conventional treatments. By contrast, the introduction of such therapies has led to the appearance of a new toxicity profile. The possible correlation between the onset of irAEs and the efficacy of the ICI drug has long been studied: it has been hypothesized that patients who show a better anticancer response, following treatment with ICI, have at the same time a greater probability of autoimmune toxicity [17]. This would seem to be caused by a more competent and reactive immune system, which stops the evolution of tumor pathology but also leads to an increase in responses against self [18]. Although several studies seem to confirm this relationship, there are still several disputes: the pathogenesis of these manifestations is not entirely clear, nor how the site, severity, timing of onset, and management of irAEs can influence the effectiveness of ICI drugs. Unfortunately, the research on biomarkers of response or toxicities is still an open field in immuno-oncology and we still do not know why certain cancer patients did not develop irAEs, and thus future research could be focused on the identification of toxicity biomarkers (sputum, microbiota). The only weapon currently available to avoid the occurrence of serious events is early interception, which is only possible through the knowledge of these manifestations. Close surveillance of patients being treated with immunosuppressive drugs allows a clinician to make a timely diagnosis when the adverse event occurs in the initial phase, allowing early treatment and avoiding progression to more advanced stages, which could lead to temporary or permanent suspension of the anticancer drug itself.

Potentially, irAEs can affect any organ. Adverse events reported more frequently include cutaneous, gastrointestinal, hepatic, endocrine, pulmonary, renal, rheumatological, neurological, cardiovascular, pericardial, and ophthalmological toxicity. These manifestations, in addition to being widely described in the literature, are reported in the guidelines "Management of Immunotherapy Toxicity" (Aimo, Italian Association of Medical Oncology), in which the correct approach for each individual toxicity is examined on the basis of the severity of these events. With regard to adverse events affecting the oral mucous membranes, there are not many studies that analyze their characteristics and methods of onset. It is possible that these events are underestimated as they are often mild or moderate and no symptomatology is associated with them. To intercept them, dedicated and accurate oral investigation is therefore necessary. For cancer patients planning to initiate immunotherapy, oral healthcare providers should be aware of any existing immune-mediated conditions. ICI therapy may exacerbate pre-existing immune-mediated conditions, and it is important to be able to distinguish a de novo irAE from exacerbation of an underlying disease process [19].

In addition, the present study showed an important correlation between skin changes and oral manifestations. Since dermatologic toxicities are among those with the highest incidence [20], it is necessary to deepen the investigations regarding irAEs involving the oral mucosa and introduce these manifestations to the guidelines for the management of immunotherapy toxicities, thereby providing clinicians with an important aid for their interception and treatment and improving the quality of life of the patient being treated with ICI drugs [21].

5. Conclusions

From the analysis of the literature and our clinical experience, the correlation between therapy with ICI drugs and the onset of irAEs appears certain. This work aims to propose clinical–diagnostic recommendations for general dentists and oral pathologists who are faced with certain situations. In the context of the management of oral lesions, pharmacological treatments (topical steroid therapies, hyaluronic acid drugs, and nystatin in the case of candidiasis) aiming to reduce the inflammatory state and burning symptomatology are useful; moreover, laser photobiomodulation treatments to prevent the onset of lesions and burning symptoms are considered to be a valid therapeutic choice [22]. For whitish lesions, it is possible to exclude potentially malignant lesions using cold-blade or laser-assisted biopsy and consequent histological examinations [23–25], although today, no onset of oral carcinomas following immunological therapy has been reported in the literature. In general, patients with a history of an oral irAE must be closely followed by their oral healthcare provider for flare-ups, recurrence, and/or new oral toxicities. In our study, considering the small sample size and the low grade of toxicities reported, we cannot evaluate the association between oral toxicity and tumor response because the patients in the present paper are all on a follow-up regimen, and a definitive evaluation after such a short period of therapy is still impossible. Further research is needed to elucidate the mechanism of these oral toxicities as well as the role they may play in predicting the antitumor efficacy of ICIs [26].

Author Contributions: Conceptualization, U.R. and A.P.; methodology, D.P. and G.P.; software, R.F.; validation, G.T., F.R. and A.C.; formal analysis, D.P.; investigation, G.P.; resources, A.B.; data curation, R.F.; writing—original draft preparation, D.P.; writing—review and editing, G.P.; visualization, U.R.; supervision, A.P.; project administration, G.T. All authors have read and agreed to the published version of the manuscript.

Funding: This research received no external funding.

Institutional Review Board Statement: The study was conducted according to the guidelines of the Declaration of Helsinki and approved by the Institutional Review Board of the Department of Oral and Maxillofacial Sciences of Sapienza University of Rome and was registered with the protocol number 001393 on 22 December 2020.

Informed Consent Statement: Informed consent was obtained from all subjects involved in the study.

Data Availability Statement: Data can be obtained from the corresponding author.

Conflicts of Interest: The authors declare no conflict of interest.

References

1. Yang, Y. Cancer immunotherapy: Harnessing the immune system to battle cancer. *J. Clin. Investig.* **2015**, *125*, 3335–3337. [CrossRef]
2. Finn, O.J. Immuno-oncology: Understanding the function and dysfunction of the immune system in cancer. *Ann. Oncol.* **2012**, *23* (Suppl. S8), viii6–viii9. [CrossRef] [PubMed]
3. Roberto, M.; Botticelli, A.; Cecere, F.; Cognetti, F.; Giusti, R.; Gelibter, A.; Lugini, A.; Nelli, F.; Nuti, M.; Santini, D.; et al. L'immunoterapia nel tumore del polmone non a piccole cellule: Ritorno al futuro [Immunotherapy in non-small cell lung cancer patients: Back to the future]. *Recenti Prog. Med.* **2019**, *110*, 587–593. (In Italian) [CrossRef] [PubMed]
4. Lipson, E.J.; Drake, C.G. Ipilimumab: An anti-CTLA-4 antibody for metastatic melanoma. *Clin. Cancer Res.* **2011**, *17*, 6958–6962. [CrossRef]
5. Topalian, S.L.; Hodi, F.S.; Brahmer, J.R.; Gettinger, S.N.; Smith, D.C.; McDermott, D.F.; Powderly, J.D.; Carvajal, R.D.; Sosman, J.A.; Atkins, M.B.; et al. Safety, activity, and immune correlates of anti-PD-1 antibody in cancer. *N. Engl. J. Med.* **2012**, *366*, 2443–2454. [CrossRef] [PubMed]
6. Postow, M.A.; Sidlow, R.; Hellmann, M.D. Immune-related adverse events associated with immune checkpoint blockade. *N. Engl. J. Med.* **2018**, *378*, 158–168. [CrossRef]
7. Dolladille, C.; Ederhy, S.; Sassier, M.; Cautela, J.; Thuny, F.; Cohen, A.A.; Fedrizzi, S.; Chrétien, B.; Da-Silva, A.; Plane, A.F.; et al. Immune Checkpoint Inhibitor Rechallenge After Immune-Related Adverse Events in Patients with Cancer. *JAMA Oncol.* **2020**, *6*, 865–871. [CrossRef]
8. Boutros, C.; Tarhini, A.; Routier, E.; Lambotte, O.; Ladurie, F.L.; Carbonnel, F.; Izzeddine, H.; Marabelle, A.; Champiat, S.; Berdelou, A.; et al. Safety profiles of anti-CTLA4 and anti-PD-1 antibodies alone and in combination. *Nat. Rev. Clin. Oncol.* **2016**, *13*, 473–486. [CrossRef]
9. Yang, W.; Li, S.; Yang, Q. Risk of dermatologic and mucosal adverse events associated with PD-1/PD-L1 inhibitors in cancer patients: A meta-analysis of randomized controlled trials. *Medicine* **2019**, *98*, e15731. [CrossRef]
10. Dika, E.; Lambertini, M.; Gouveia, B.; Mussi, M.; Marcelli, E.; Campione, E.; Gurioli, C.; Melotti, B.; Alessandrini, A.; Ribero, S. Oral Manifestations in Melanoma Patients Treated with Target or Immunomodulatory Therapies. *J. Clin. Med.* **2021**, *10*, 1283. [CrossRef]
11. Sibaud, V.; Eid, C.; Belum, V.R.; Combemale, P.; Barres, B.; Lamant, L.; Mourey, L.; Gomez-Roca, C.; Estilo, C.L.; Motzer, R.; et al. Oral lichenoid reactions associated with anti-PD-1/PD-L1 therapies: Clinicopathological findings. *J. Eur. Acad. Dermatol. Venereol.* **2017**, *31*, e464–e469. [CrossRef] [PubMed]
12. Fazer, C.; Price, K.A. Management of immune-related dermatitis and mucositis associated with pembrolizumab in metastatic human papillomavirus-associated squamous cell carcinoma of the oropharynx. *JCO Oncol. Pract.* **2020**, *16* (Suppl. S2), 20s–24s. [CrossRef] [PubMed]

13. Acero Brand, F.Z.; Suter, N.; Adam, J.P.; Faulques, B.; Maietta, A.; Soulières, D.; Blais, N. Severe immune mucositis and esophagitis in metastatic squamous carcinoma of the larynx associated with pembrolizumab. *J. Immunother. Cancer* **2018**, *6*, 22. [CrossRef]
14. Lacouture, M.; Sibaud, V. Toxic Side Effects of Targeted Therapies and Immunotherapies Affecting the Skin, Oral Mucosa, Hair, and Nails. *Am. J. Clin. Dermatol.* **2018**, *19* (Suppl. S1), 31–39. [CrossRef] [PubMed]
15. Warner, B.M.; Baer, A.N.; Lipson, E.J.; Allen, C.; Hinrichs, C.; Rajan, A.; Pelayo, E.; Beach, M.; Gulley, J.L.; Madan, R.A.; et al. Sicca Syndrome Associated with Immune Checkpoint Inhibitor Therapy. *Oncologist* **2019**, *24*, 1259–1269. [CrossRef] [PubMed]
16. Del Vecchio, A.; Palaia, G.; Grassotti, B.; Tenore, G.; Ciolfi, C.; Podda, G.; Impellizzeri, A.; Mohsen, A.; Galluccio, G.; Romeo, U. Effects of laser photobiomodulation in the management of oral lichen planus: A literature review. *Clin. Ther.* **2021**, *172*, 467–483. [CrossRef]
17. Zumelzu, C.; Alexandre, M.; Le Roux, C.; Weber, P.; Guyot, A.; Levy, A.; Aucouturier, F.; Mignot-Grootenboer, S.; Caux, F.; Maubec, E.; et al. Mucous membrane pemphigoid, bullous pemphigoid, and anti-programmed death-1/programmed death-ligand 1: A case report of an elderly woman with mucous membrane pemphigoid developing after pembrolizumab therapy for metastatic melanoma and review of the literature. *Front. Med.* **2018**, *5*, 268.
18. Dascorresponding, S.; Johnson, D.B. Immune-related adverse events and anti-tumor efficacy of immune checkpoint inhibitors. *J. Immunother. Cancer* **2019**, *7*, 306. [CrossRef]
19. Klein, B.A.; Shazib, M.A.; Villa, A.; de Abreu Alves, F.; Vacharotayangul, P.; Sonis, S.; Fedele, S.; Treister, N.S. Immune checkpoint inhibitors in cancer therapy: Review of orofacial adverse events and role of the oral healthcare provider. *Front. Oral Health* **2022**, *3*, 968157. [CrossRef]
20. Riudavets, M.; Barba, A.; Maroto, P.; Sullivan, I.G.; Anguera, G.; Paez, D.; del Carpio, L.; Callejo, A.; Gonzalez Blanco, C.; Garcia Planellas, E.; et al. Correlation between immune-related adverse events (irAEs) and efficacy in patients with solid tumors treated with immune-checkpoints inhibitors (ICIs). *J. Clin. Oncol.* **2018**, *36* (Suppl. S15), 3064. [CrossRef]
21. Tattersall, I.W.; Leventhal, J.S. Cutaneous Toxicities of Immune Checkpoint Inhibitors: The Role of the Dermatologist. *Yale J. Biol. Med.* **2020**, *93*, 123–132. [PubMed]
22. Del Vecchio, A.; Tenore, G.; Luzi, M.C.; Palaia, G.; Mohsen, A.; Pergolini, D.; Romeo, U. Laser Photobiomodulation (PBM)-A Possible New Frontier for the Treatment of Oral Cancer: A Review of In Vitro and In Vivo Studies. *Healthcare* **2021**, *9*, 134. [CrossRef] [PubMed]
23. Palaia, G.; D'Alessandro, L.; Pergolini, D.; Carletti, R.; Di Gioia, C.R.T.; Romeo, U. In vivo clinical and histological thermal effect of a 445 nm diode laser on oral soft tissues during a biopsy. *J. Oral Sci.* **2021**, *63*, 280–282. [CrossRef]
24. Palaia, G.; Renzi, F.; Pergolini, D.; Del Vecchio, A.; Visca, P.; Tenore, G.; Romeo, U. Histological Ex Vivo Evaluation of the Suitability of a 976 nm Diode Laser in Oral Soft Tissue Biopsies. *Int. J. Dent.* **2021**, *2021*, 6658268. [CrossRef]
25. Pergolini, D.; Del Vecchio, A.; Mohsen, M.; Tenore, G. Inter-incisor ossifying fibroma removal by diode laser. *Dent. Cadmos* **2020**, *88*, 266–268. [CrossRef]
26. Klein, B.A.; Alves, F.A.; de Santana Rodrigues Velho, J.; Vacharotayangul, P.; Hanna, G.J.; LeBoeuf, N.R.; Shazib, M.A.; Villa, A.; Woo, S.B.; Sroussi, H.; et al. Oral manifestations of immune-related adverse events in cancer patients treated with immune checkpoint inhibitors. *Oral Dis.* **2022**, *28*, 9–22. [CrossRef]

applied sciences

MDPI

Article

Oral Pathologies in Migrants: The Experience of the "PROTECT" Project in 3023 Patients

Gaspare Palaia [1,*], Maurizio Bossù [1], Nicola Pranno [1,*], Massimo Ralli [2], Alice Bruscolini [2], Mauro Capocci [1], Gianfranco Gaimari [1], Federica Macali [1], Denise Corridore [1], Luisa Cotticelli [1], Francesco Covello [1], Michele Ahmed Antonio Karaboue [1], Stefano Di Carlo [1], Umberto Romeo [1], Livia Ottolenghi [1] and Antonella Polimeni [1]

1 Department of Oral and Maxillofacial Sciences, Sapienza University of Rome, 00161 Rome, Italy
2 Department of Sense Organs, Sapienza University of Rome, 00185 Roma, Italy
* Correspondence: gaspare.palaia@uniroma1.it (G.P.); nicola.pranno@uniroma1.it (N.P.)

Abstract: Introduction: The number of people with migrant status living in Europe is proliferating. Most of the refugees in Italy come from war zones, and many of them denounce having been victims of persecutory acts in their country of origin. Highly cultured migrant populations have shown better results and oral health behaviour than those who were poorly cultured. The PROTECT project aimed to build a network for the early management of head and neck pathologies among refugees and migrants, promoting the dissemination of correct information about the prevention and treatment of these pathologies. Materials and methods: A national cross-sectional study among the refugees and migrant population in the Lazio region, Italy, from February 2018 to September 2021 was performed. The oral health of 3023 participants was investigated within a network of 56 reception centres and cultural associations. Data collected via an oral health survey questionnaire gathered information on participants' demographic factors, migration status and dental behaviours and clinical examinations of the participants with the help of mouth mirrors, periodontal probes and artificial light. Results: The mean age was 31.6 ± 13.1, and among all the subjects, 2058 were male (68.1%) and 965 were women (41.9%). Most participants were born in Nigeria, followed by Bangladesh, Pakistan, Somalia, Mali and Senegal. The overall oral pain prevalence was 48.2%. The prevalence of patients claiming poor oral hygiene was 32.4%; 36.2% of the subjects consumed high amounts of sugar; and 26.7% saw their dentist for a check-up in the last year. At the clinical examination, 68.9% of patients had caries experience (considering decay of deciduous teeth, and caries of permanent teeth and teeth with fillings), with 32.2% showing pulpal involvement. Low levels of oral hygiene were also found at the clinical examination, with 46.5% of patients presenting plaque and calculus. The trend of the DMFT index was found to be 5.41. Good periodontal health (CPI = 0) was present in approximately 33.5% of patients. The CPI = 1 index reported bleeding from gingivitis in 37% of patients; tartar was found in 27% of patients (CPI = 2). The percentage of patients with CPI = 3 was 3.6%. Just over half (52.2%) of the migrants examined had malocclusion, and only 0.7% had a malocclusion in treatment. Conclusions: The goal to be achieved is to develop education and prevention programs for head–neck diseases, and perhaps even more. The first step towards this goal can be removing the obstacles migrants encounter in accessing health care.

Keywords: migrants; vulnerability; assistance; hospitality; minors; dentistry; inclusion

check for updates

Citation: Palaia, G.; Bossù, M.; Pranno, N.; Ralli, M.; Bruscolini, A.; Capocci, M.; Gaimari, G.; Macali, F.; Corridore, D.; Cotticelli, L.; et al. Oral Pathologies in Migrants: The Experience of the "PROTECT" Project in 3023 Patients. *Appl. Sci.* **2022**, *12*, 12621. https://doi.org/10.3390/app122412621

Academic Editor: Gaetano Isola

Received: 2 November 2022
Accepted: 6 December 2022
Published: 9 December 2022

1. Introduction

The number of people with migrant status living in Europe is proliferating. According to the United Nations High Commissioner for Refugees (UNCHR), the number of refugees reached 84 million worldwide in 2021. Most of the refugees in Italy come from war zones, and many of them denounce having been victims of persecutory acts in their country of origin [1]. The reasons that lead people to flee are: conflict, COVID-19, poverty, political instability and increased globalisation [2]. Over the years, it has been observed that migrants from middle- and low-income countries migrate to high-income countries such as the USA, Canada, Australia, and those in Europe. Several factors help to define migrants as vulnerable: health risks before, during, and after migration; a disease profile different from that of the host country population; and barriers to accessing health services in host countries [3]. In the same way, it has become increasingly evident how culture, beliefs, and ethnic habits can influence oral health practices in migrants [4]. Highly cultured migrant populations have shown better results and oral health behaviour than those poorly cultured [5]. Moreover, language, social, and economic barriers make access to care in public health institutions difficult. Because of the high costs for treatment in private facilities, there is often a considerable incidence of dental and oral disease in migrant patients. Bad oral health could affect the quality of life, as it could interfere with daily activities such as eating and talking. In addition, untreated oral health problems may pose severe risks to the health of older people, including malnutrition, strokes, heart disease, pneumonia, and oral and pharyngeal cancers [6]. For all these reasons, new intervention strategies, such as the PROTECT (Patologie del distRettO TEsta-Collo nei migranTi. Dalla formazione degli operatori alla diagnosi precoce e presa in carico del paziente: Network Odontoiatrico • Oftalmologico • Otorinolaringoiatrico • Maxillo-Facciale) project, were created to protect the health of vulnerable applicants and holders of international protection. The PROTECT project was funded by the Asylum, Migration, and Integration Fund (FAMI) 2014–2020 of the Ministry of the Interior and co-financed by the European Union. On the one hand, PROTECT's activities are concerned with screening actions directly at reception centres or organisations that deal with refugees and migrants, and subsequently taking early charge of cases with pathologies to be treated. On the other hand, training and awareness-raising activities both for healthcare personnel (including those in training) and for the personnel of reception centres are also areas of concern.

Patients with treated vulnerabilities were minors, unaccompanied foreign minors, disabled people, pregnant women, single parents with minor children, victims of trafficking, people with illnesses or mental disorders, and people who have been tortured, raped or other forms of psychological, physical, and sexual violence.

2. Materials and Methods

The PROTECT project aimed to build a network for early management of head and neck pathologies among refugees and migrants, promoting the dissemination of correct information about the prevention and treatment of these pathologies. A national cross-sectional study among the refugees and migrant population present in the Lazio region, Italy, from February 2018 to September 2021 was performed. The oral health of 3023 participants was investigated within a network of 56 reception centres and cultural associations. The "PROTECT" project was approved by the Department of Oral and Maxillofacial Sciences, Sapienza, University of Rome (Protocol identifying number: 0000839 on 2 October 2018). The protocol was in accordance with the 1975 Declaration of Helsinki on medical protocols and ethics and its later amendments.

2.1. Data Collection

Data collected via an oral health survey questionnaire gathered information on participants' demographic factors, migration status, and dental behaviours; clinical examinations of the participants were performed with the help of mouth mirrors, periodontal probes, and artificial light. The interviews and oral examinations were performed by trained and calibrated dentists of the Department of Oral and Maxillofacial Sciences, Sapienza, University of Rome.

The screenings took place at both the premises of the reception centres and the premises of the Policlinico Umberto I, a mobile vehicle was used specifically for visits. The recording of the pathologies found was carried out through the use a screening folder, with red or yellow codes, depending on the severity of the disease.

2.2. Questionnaire Information

The questionnaire was administered individually through interviews including the following areas of interest:

1. Sociodemographic characteristics: first name; surname; gender; date of birth; age; country of origin.
2. Dental health-related behaviours: have you ever had an injury in the head and neck region? (no/yes); Do you brush your teeth at least twice a day? (no/yes); Do you eat a lot of foods that are highest in sugar? (no/yes); Do you have pain in your mouth right now? (no/yes); Have you visited the dentist in the past year? (no/yes).

2.3. Clinical Examination

The subjects were examined in the cultural associations, in the reception centres, or in specially equipped motor vehicles for evaluation of the head and neck region. Three examiners were calibrated during the examinations on the first 50 patients and generated a good intra-class correlation coefficient (Kappa = 0.84, range: 0.75–0.91). The dependent variables investigated were: the sum of decayed, missing, and filled permanent teeth (the DMFT index); periodontal disease (periodontal health, gingivitis, suspected periodontitis); the cumulative periodontal index (CPI); the presence of oral mucosal lesion (yes/no); the presence of malocclusion (yes/no).

The intraoral clinical evaluation was performed on each patient using a sealed sterilised kit consisting of a mirror and a disposable probe, and cotton rolls to dry the dental surfaces, with the patient lying down under a good light source (natural light and fluorescent lamp). Second-level examinations, such as orthopanoramic radiography, were requested for the diagnosis of carious lesions (not detectable in the extraoral clinical examination) and any anomalies in the number and position of the dental elements, and then carried out in the relevant competent departments at the Umberto I Polyclinic. The data collected were recorded in the patient's medical record.

2.4. Data Analysis

Data were evaluated using standard statistical analysis software (version 20.0, Statistical Package for the Social Sciences, IBM Corporation, Armonk, NY, USA). A database was created using Excel (Microsoft, Redmond, WA, USA). Descriptive statistics, including mean $\pm$ SD values and percentages were calculated for each variable.

The following continuous and categorical variables were explored: age, gender, nationality, history of trauma to the oral region, oral pain, inadequate oral hygiene, presence of oral neoformations, presence of pulp exposure, presence of fistulas, presence of abscess, presence of root surface exposure, the plaque index, and the DMFT scores. Multiple linear regression analysis was performed to ascertain the effects of independent variables on the DMFT scores. Statistical significance was set at $p \leq 0.05$.

3. Results

A total of 3023 participants were examined and interviewed. The mean age was 31.6 ± 13.1, and among all the subjects, 2058 were male (68.1%) and 965 were women (41.9%). Most participants were born in Nigeria, followed by Bangladesh, Pakistan, Somalia, Mali and Senegal (Figure 1). The overall oral pain prevalence was 48.2%. The prevalence of patients who claimed poor oral hygiene was 32.4%; 36.2% of the subjects consumed high sugar; and 26.7% saw their dentist for a check-up in the last year. At the clinical examination, 68.9% of patients had caries experience (considering decay of deciduous teeth, and caries of permanent teeth and teeth with fillings), with 32.2% showing pulpal involvement. Low levels of oral hygiene were also found at the clinical examination, with 46.5% of patients presenting plaque and calculus. The trend of the DMFT index, which expresses the mean value of dental pathology, was found to be 5.41, determined by the carious component (Decay: 2.21), the lack of permanent teeth (Missing: 1.78), and the teeth with fillings (Filling: 1.42) (Figure 2). The multiple regression model statistically significantly predicted an increase in DMFT scores correlated with an increase in age (p-value = 0.003) (Table 1). Periodontal health has been expressed with the CPI (Community Periodontal Index) when going to evaluate gingival bleeding, calculus, and periodontal pockets (3 mm < PD < 6 mm). Good periodontal health (CPI = 0) was present in approximately 33.5% of patients. The CPI = 1 index reported bleeding from gingivitis in 37% of patients; tartar was found in 27% of patients (CPI = 2). The percentage of patients with CPI = 3 was 3.6%. Just over half (52.2%) of the migrants examined had malocclusion, and only 0.7% had a malocclusion in treatment.

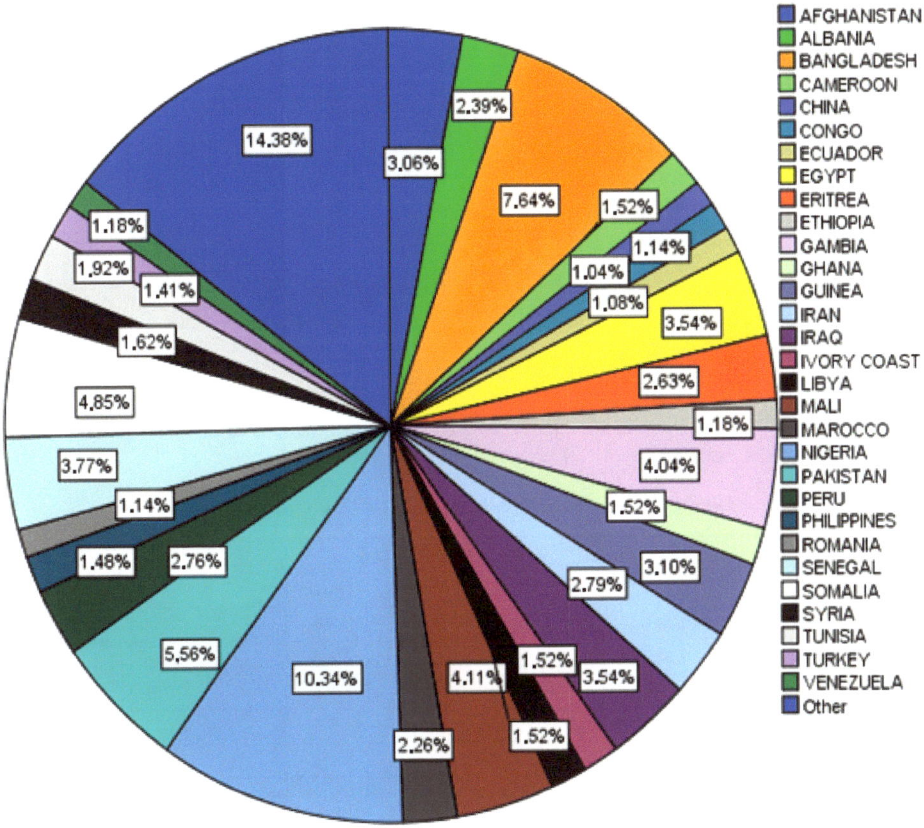

Figure 1. Percentage of subjects coming from different countries.

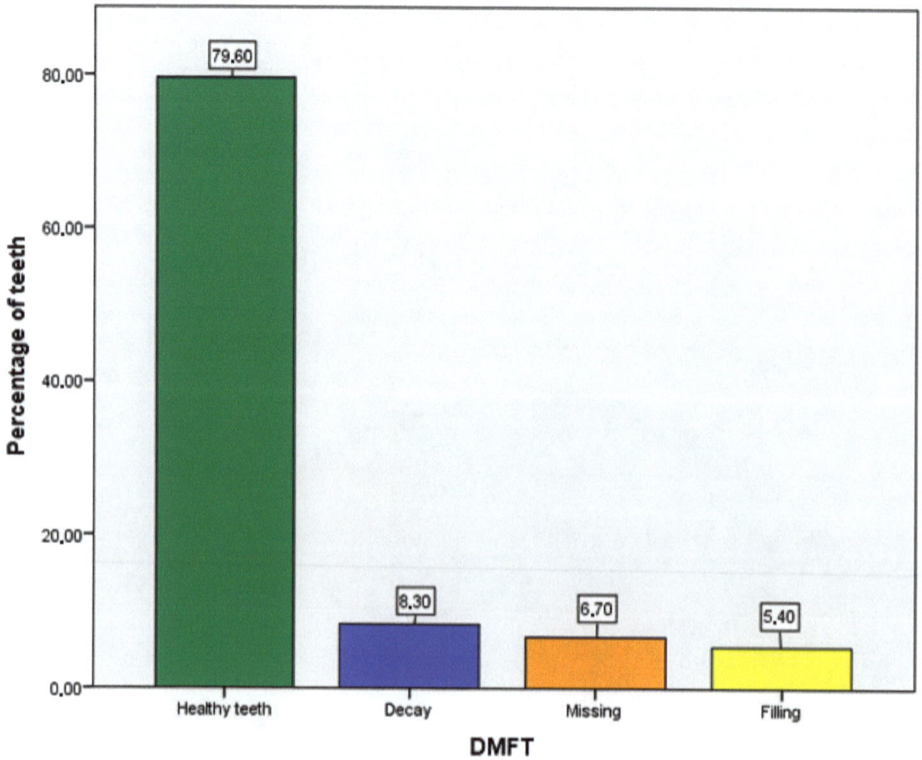

Figure 2. Percentage of the healthy, decayed, missing, and filled teeth on a total of 3023 patients evaluated.

Table 1. Multiple linear regression results for DMFT scores.

DMFT Scores	β Coefficient	95% C.I. for β		SE β	*p*-Value
		Lower	Upper		
Age	0.090	0.035	0.186	0.153	0.003
Gender	0.360	−0.886	2.645	0.053	0.552
Presence of oral pain	0.730	0.154	1.405	0.221	0.363
Inadequate oral hygiene	0.367	0.106	0.607	0.359	0.103
Plaque index	0.227	−0.004	0.458	0.182	0.084

Note: β = unstandardised regression coefficient; SE β = standard error of coefficient; C.I. = confidence interval.

4. Discussion

In recent years, the prevalence of caries and periodontal disorders in industrialised countries has decreased in adults and children due to the use of preventive measures. However, the data collected by our screening did not show the same situation between migrants and asylum seekers. Data collected from the literature show that the migrant status may be related to a higher prevalence of dental caries [7–13] and a lower periodontal condition [14–16]. Difficulties ensuring access to basic needs for migrant populations, such as food, water, sanitation, and health care, have been documented in the literature [17]. Research conducted in Colombia in 2019 [18] has suggested and demonstrated that oral health is considered a secondary problem for vulnerable people who have to change their lifestyle and adapt to another culture. However, several studies [19,20] show that acculturation, defined as "people's lifestyle and behavioural changes when they move from one culture and adapt to another culture, usually because of immigration", affects oral health. Refugees and asylum seekers have difficulty gaining access to oral health care [21]. The health system, society, and personal behaviour in the area of oral health determine access to health care services. The health policy of the host country is a crucial element in determining access to oral health care. The highest prevalence of caries in the group of migrants in the research of Ferrazzano et al. [22] was associated with difficulties in gaining access to health services. Al-Ani et al. (Germany) [23] noted in their study that the incidence of oral disorders is greater in people with low socioeconomic status, such as asylum seekers who find themselves having to overcome many difficulties to start their lives again in a foreign country. The study by Salim et al. [24] reported that pain was the most commonly presented disorder in 444 patients, stating that more than half of the dental treatments carried out were extractions, 74.1% of which were due to caries. Similarly, pain was a common symptom reported by participants in the research of Saadeh et al. [25], and more than half of the participants reported needing dental care, but not being able to receive it given their low socioeconomic status. Research by Zinah et al. [26] shows that periodontal disease and dental caries are the most frequently assessed study conditions. Salim et al. [27] analysed oral hygiene practices among refugees, noting that most brushed their teeth less than twice a day, thus having poor oral hygiene. The oral hygiene practices and behaviours of parents inevitably directly influence the oral health of children. Literacy and parents' education regarding good oral health practices may reduce the dmft/DMFT value for their children [28]. A study conducted in Canada between 2013 and 2016 by Moreau et al. [29] found that approximately 60% of refugee children had never visited a dentist before while most Canadian counterparts had seen a dentist in the last year. In our study, the figure is even more worrying as only a small percentage had performed a visit in the last year. Finding higher dmft/DMFT values and more common anterior cross-bites in refugee children, no differences were noted from other types of malocclusion. This project shows how the problem of the poor oral health of migrants exists and is widely documented in the literature. As oral health strongly influences the quality of life, it is necessary to make oral health services more easily accessible to persons in a vulnerable state under the status of migrant [30]. The screening action is carried out directly in reception centres and/or organisations that deal with refugees and migrants in the Lazio region, making it possible to take a snapshot of oral health at the time of legal acceptance of migrant patients and to highlight a request for necessary care, as also found in the literature. More than half of the subjects that were examined experienced caries with painful symptoms at the time of the visit as a result of an impossibility of treatment in the countries of origin and difficulty in accessing the National Health Service due not only to language barriers but also to a delay in identifying the legal recognition that facilitates access. The incidence of caries in our sample is also linked to poor hygiene, as well as to a high intake of sugar, present in approximately half of the sample, although periodontal problems are not one of the prevalent problems. In line with the literature, the data show that malocclusions, detected in half of the sample, are almost entirely not treated, due to socioeconomic status and access difficulties.

The need for a project such as PROTECT is to improve the quality of life of these patients after the resolution of the discomforts, often linked to acute painful symptoms of an inflammatory/infectious nature, which, without early intervention, could lead to even more severe problems (e.g., neoplasms).

In the present study, it is important to consider the high DMFT score (5.4), likely due to a "westernization" of the diet favouring caries. Furthermore, worrying is the fact that the percentage of missing teeth (M) or still decayed (D) is higher than the treated teeth (F), confirming the need for a campaign to prevent and treat carious pathology in this category of vulnerable people [31–33]. Moreover, this score contrasts with the values observed in the epidemiological study of Vano et al. [34], in which the authors reported in a cohort of adult Italian patients an approximate DMFT score of 4. This difference may be associated with the absence of healthy structures to treat and prevent oral pathologies in the country of origin.

5. Conclusions

PROTECT has been an innovative project of fundamental importance for the beneficiaries, having the ability to involve approximately 4000 people, including operators of the centres, patients and clinicians. In this project, they found a point of reference previously unthinkable. Knowledge, prevention, and management of head–neck diseases (which are often painful and disabling in these patients who, for contingent reasons related to their condition as migrants, are usually neglected) has led to an improvement in the quality of life of these patients. In addition, the PROTECT project not only helped multiple underprivileged people but also provided an understanding of the pathologies that most commonly occur among migrants. This will make it possible in the future to create even more targeted and specific and, therefore, more efficient programs.

So, we can conclude that:

- Dental pathologies (especially untreated carious lesions or malocclusions) in migrants are unfortunately widespread and often neglected.
- Prevention and early intervention avoid hospitalisations and worsening problems.
- PROTECT has been a point of reference for patients and operators of the reception centres.

Author Contributions: Conceptualisation, G.P. and M.B.; data curation, G.P., N.P. and M.R.; formal analysis, M.C.; investigation, D.C., L.C. and F.C.; methodology, M.A.A.K. and S.D.C.; project administration, L.O. and A.P.; resources, L.O. and A.P.; software, N.P. and M.C.; supervision, U.R., L.O. and A.P.; validation, D.C. and F.C.; visualisation, G.P., M.R. and U.R.; writing—original draft, A.B., G.G. and F.M.; writing—review and editing, G.P., M.B. and L.C. All authors have read and agreed to the published version of the manuscript.

Funding: This research received external funding from the Asylum, Migration and Integration Fund (AMIF) of the European Union, according to the code PROG-2010.

Institutional Review Board Statement: The "PROTECT" project was approved by the Department of Oral and Maxillofacial Sciences, Sapienza, University of Rome (Protocol identifying number: 0000839 on 2 October 2018). The protocol was in accordance with the 1975 Declaration of Helsinki on medical protocols and ethics and its later amendments.

Informed Consent Statement: Informed consent was obtained from all subjects involved in the study.

Data Availability Statement: The datasets used and/or analysed during the current study are available from the corresponding author upon reasonable request.

Conflicts of Interest: The authors declare no conflict of interest.

References

1. Pinchi, V.; Focardi, M.; Pradella, F.; Grifoni, R.; Palandri, M.; Norelli, G.A. Day to day issues in the forensic identification practice related to illegal immigration in Italy. *J. Forensic. Odontostomatol.* **2017**, *35*, 157–165. [PubMed]
2. Dao, T.H.; Docquier, F.; Maurel, M.; Schaus, P. Global migration in the twentieth and twenty-first centuries: The unstoppable force of demography. *Rev. World Econ.* **2021**, *157*, 417–449. [CrossRef]
3. Lauritano, D.; Moreo, G.; Carinci, F.; Campanella, V.; Della Vella, F.; Petruzzi, M. Oral Health Status among Migrants from Middle- and Low-Income Countries to Europe: A Systematic Review. *Int. J. Environ. Res. Public Health* **2021**, *18*, 12203. [CrossRef]
4. Dahlan, R.; Badri, P.; Saltaji, H.; Amin, M. Impact of acculturation on oral health among immigrants and ethnic minorities: A systematic review. *PLoS ONE* **2019**, *14*, e0212891. [CrossRef] [PubMed]
5. Pabbla, A.; Duijster, D.; Grasveld, A.; Sekundo, C.; Agyemang, C.; van der Heijden, G. Oral Health Status, Oral Health Behaviours and Oral Health Care Utilisation among Migrants Residing in Europe: A Systematic Review. *J. Immigr. Minor. Health.* **2021**, *23*, 373–388. [CrossRef]
6. Hoang, H.; Feike, S.; Lynden, T.; Barnett, T.; Crocombe, L. Oral health needs of older migrants with culturally and linguistically diverse backgrounds in developed countries: A systematic review. *Australas. J. Ageing* **2020**, *39*, 193–208. [CrossRef]
7. Delgado-Angulo, E.K.; Marcenes, W.; Harding, S.; Bernabé, E. Ethnicity, migration status and dental caries experience among adults in East London. *Community Dent. Oral Epidemiol.* **2018**, *46*, 392–399. [CrossRef]

8. Aarabi, G.; Reissmann, D.R.; Seedorf, U.; Becher, H.; Heydecke, G.; Kofahl, C. Oral health and access to dental care—A comparison of elderly migrants and non-migrants in Germany. *Ethn. Health* **2018**, *23*, 703–717. [CrossRef]
9. Erdsiek, F.; Waury, D.; Brzoska, P. Oral health behaviour in migrant and non-migrant adults in Germany: The utilization of regular dental check-ups. *BMC Oral Health* **2017**, *17*, 84. [CrossRef]
10. Zero, D.T.; Fontana, M.; Martínez-Mier, E.A.; Ferreira-Zandoná, A.; Ando, M.; González-Cabezas, C.; Bayne, S. The biology, prevention, diagnosis and treatment of dental caries: Scientific advances in the United States. *J. Am. Dent. Assoc.* **2009**, *140* (Suppl. S1), 25S–34S. [CrossRef]
11. Ponce-Gonzalez, I.; Cheadle, A.; Aisenberg, G.; Cantrell, L.F. Improving oral health in migrant and underserved populations: Evaluation of an interactive, community-based oral health education program in Washington state. *BMC Oral Health* **2019**, *19*, 30. [CrossRef] [PubMed]
12. Gibbs, L.; Waters, E.; de Silva, A.; Riggs, E.; Moore, L.; Armit, C.; Johnson, B.; Morris, M.; Calache, H.; Gussy, M.; et al. An exploratory trial implementing a community-based child oral health promotion intervention for Australian families from refugee and migrant backgrounds: A protocol paper for Teeth Tales. *BMJ Open* **2014**, *4*, e004260. [CrossRef] [PubMed]
13. Spinler, K.; Weil, M.T.; Valdez, R.; Walther, C.; Dingoyan, D.; Seedorf, U.; Heydecke, G.; Lieske, B.; Kofahl, C.; Aarabi, G. Mundgesundheitskompetenz von Menschen mit Migrationshintergrund—Erste Auswertungen der MuMi-Studie [Oral health literacy of persons with migration background-first results of the MuMi study]. *Bundesgesundheitsblatt Gesundh. Gesundh.* **2021**, *64*, 977–985. (In German) [CrossRef]
14. Traisuwan, W. Oral health status and behaviors of pregnant migrant workers in Bangkok, Thailand: A cross-sectional study. *BMC Oral Health* **2021**, *21*, 379. [CrossRef] [PubMed]
15. Hagenfeld, D.; Zimmermann, H.; Korb, K.; El-Sayed, N.; Fricke, J.; Greiser, K.H.; Kühnisch, J.; Linseisen, J.; Meisinger, C.; Schmitter, M.; et al. Periodontal Health and Use of Oral Health Services: A Comparison of Germans and Two Migrant Groups. *Int. J. Environ. Res. Public Health* **2019**, *16*, 3000. [CrossRef]
16. Dahlan, R.; Ghazal, E.; Saltaji, H.; Salami, B.; Amin, M. Impact of social support on oral health among immigrants and ethnic minorities: A systematic review. *PLoS ONE* **2019**, *14*, e0218678. [CrossRef]
17. Abbas, M.; Aloudat, T.; Bartolomei, J.; Carballo, M.; Durieux-Paillard, S.; Gabus, L.; Jablonka, A.; Jackson, Y.; Kaojaroen, K.; Koch, D.; et al. Migrant and refugee populations: A public health and policy perspective on a continuing global crisis. *Antimicrob Resist Infect. Control* **2018**, *7*, 113. [CrossRef] [PubMed]
18. Murillo-Pedrozo, A.M.; Martínez-Herrera, E.; Ronda-Pérez, E.; Agudelo-Suárez, A.A. A Qualitative Study of the Health Perceptions in the Venezuelan Immigrant Population in Medellín (Colombia) and Its Conditioning Factors. *Int. J. Environ. Res. Public Health* **2021**, *18*, 3897. [CrossRef]
19. Crespo, E. The Importance of Oral Health in Immigrant and Refugee Children. *Children* **2019**, *6*, 102. [CrossRef]
20. Hunter-Adams, J.; Cochran, J.; Laird, L.D.; Paasche-Orlow, M.K.; Geltman, P.L. Acculturation and a Potential Relationship with Oral Health Outcomes Among Somali Refugees in Massachusetts. *J. Immigr. Minor. Health* **2018**, *20*, 351–359. [CrossRef]
21. Keboa, M.T.; Hiles, N.; Macdonald, M.E. The oral health of refugees and asylum seekers: A scoping review. *Global. Health* **2016**, *12*, 59.
22. Ferrazzano, G.F.; Cantile, T.; Sangianantoni, G.; Ingenito, A.; Rengo, S.; Alcidi, B.; Spagnuolo, G. Oral health status and Unmet Restorative Treatment Needs (UTN) in disadvantaged migrant and not migrant children in Italy. *Eur. J. Paediatr. Dent.* **2019**, *20*, 10–14. [PubMed]
23. Al-Ani, A.; Takriti, M.; Schmoeckel, J.; Alkilzy, M.; Splieth, C.H. National oral health survey on refugees in Germany 2016/2017: Caries and subsequent complications. *Clin. Oral Investig.* **2021**, *25*, 2399–2405. [CrossRef] [PubMed]
24. Salim, N.A.; Maayta, W.; ElSa'aideh, B.B. The oral health of refugees: Issues and challenges arising from a case series analysis. *Community Dent. Oral Epidemiol.* **2020**, *48*, 195–200. [CrossRef] [PubMed]
25. Saadeh, R.; Cappelli, D.; Bober-Moken, I.; Cothron, A.; de la Torre, M. Assessing Oral Health Status, Practices, and Access to Care among War-Affected Refugees Living in San Antonio, Texas. *Eur. J. Dent.* **2020**, *14*, 371–379. [CrossRef]
26. Zinah, E.; Al-Ibrahim, H.M. Oral health problems facing refugees in Europe: A scoping review. *BMC Public Health* **2021**, *21*, 1207. [CrossRef]
27. Salim, N.A.; Sawair, F.A.; Satterthwaite, J.D.; Al-Zubi, Z. Oral Hygiene Awareness, Practices and Attitudes among Syrian Refugees in Zaatari Camp and Their Impact on Oral Health Status. *Oral Health Prev. Dent.* **2021**, *19*, 689–698.
28. Reza, M.; Amin, M.S.; Sgro, A.; Abdelaziz, A.; Ito, D.; Main, P.; Azarpazhooh, A. Oral health status of immigrant and refugee children in North America: A scoping review. *J. Can. Dent. Assoc.* **2016**, *82*, g3. [PubMed]
29. Moreau, A.M.; Hennous, F.; Dabbagh, B.; Ferraz Dos Santos, B. Oral Health Status of Refugee Children in Montreal. *J. Immigr. Minor. Health* **2019**, *21*, 693–698. [CrossRef]
30. Chaffee, B.W.; Rodrigues, P.H.; Kramer, P.F.; Vítolo, M.R.; Feldens, C.A. Oral health-related quality-of-life scores differ by socioeconomic status and caries experience. *Community Dent. Oral Epidemiol.* **2017**, *45*, 216–224. [CrossRef]
31. Palaia, G.; Semprini, F.; De-Giorgio, F.; Ralli, M.; Bossù, M.; Arcangeli, A.; Ercoli, L.; Polimeni, A. Evaluation of dental demands and needs of people experiencing homelessness in the city of Rome, Italy. *Eur. Rev. Med. Pharmacol. Sci.* **2022**, *26*, 5460–5465.
32. Palaia, G.; Bruscolini, A.; Ralli, M. The Prevention and Treatment of Medical Diseases in Vulnerable Populations. *Appl. Sci.* **2022**, *12*, 10406. [CrossRef]

33. Iacoella, C.; De-Giorgio, F.; Palaia, G.; Ferraioli, M.; Arcangeli, A.; Ralli, M. Evaluation of General Health Status of Persons Living in Socio-Economically Disadvantaged Neighborhoods in a Large European Metropolitan City. *Appl. Sci.* **2022**, *12*, 7428. [CrossRef]
34. Vano, M.; Gennai, S.; Karapetsa, D.; Miceli, M.; Giuca, M.R.; Gabriele, M.; Graziani, F. The influence of educational level and oral hygiene behaviours on DMFT index and CPITN index in an adult Italian population: An epidemiological study. *Int. J. Dent. Hyg.* **2015**, *13*, 151–157. [CrossRef] [PubMed]

applied sciences

MDPI

Review

The Influence of the Menstrual Cycle and Oral Contraceptives on Knee Laxity or Anterior Cruciate Ligament Injury Risk: A Systematic Review

Juliette Moriceau [1], Amandine Fevre [1], Diego Domínguez-Balmaseda [1,2], Ángel González-de-la-Flor [1], Julia Simón-Areces [1] and Guillermo García-Pérez-de-Sevilla [1,*]

1 Faculty of Sport Sciences, Department of Physiotherapy, Universidad Europea de Madrid, 28223 Madrid, Spain
2 Masmicrobiota Group, Faculty of Health Sciences, Universidad Europea de Madrid, 28223 Madrid, Spain
* Correspondence: guillermo.garcia@universidadeuropea.es

Abstract: Women are two to four times more prone to anterior cruciate ligament (ACL) injuries than men. This raises questions about the role of the hormonal cycle in knee laxity, which may lead to increased tibial displacement and thus ACL tears. The objective was to update scientific knowledge on the influence of the menstrual cycle on knee laxity and the risk of ACL injury, with a focus on anterior tibial displacement, and on hormonal levels influenced or not by oral contraceptive use. Observational studies obtained from Pubmed, Web of Sciences and Scopus and published between 2015 and 2022 were included in this review. Studies were required to include data on menstrual cycle with/without oral contraceptives (OC) and knee laxity and/or ACL injury. A total of ten studies were selected for this systematic review. Three studies about hormone concentration and knee laxity showed an increase in estradiol during the ovulatory phase compared to the follicular phase. Of the five studies on OC, four showed a decrease in ACL laxity. Finally, four studies assessed ACL injury. The menstrual cycle appears to influence knee laxity in women. An increase in certain hormone levels was observed in the ovulatory and luteal phases when the anterior tibial translation was greater in the knee. However, based on the literature, we cannot conclude that there is a correlation between the menstrual cycle and the risk of ACL injury.

Keywords: anterior cruciate ligament; menstrual cycle; women; knee

Citation: Moriceau, J.; Fevre, A.; Domínguez-Balmaseda, D.; González-de-la-Flor, Á.; Simón-Areces, J.; García-Pérez-de-Sevilla, G. The Influence of the Menstrual Cycle and Oral Contraceptives on Knee Laxity or Anterior Cruciate Ligament Injury Risk: A Systematic Review. *Appl. Sci.* **2022**, *12*, 12627. https://doi.org/10.3390/app122412627

Academic Editors: Massimo Ralli, Gaspare Palaia and Alice Bruscolini

Received: 8 November 2022
Accepted: 7 December 2022
Published: 9 December 2022

1. Introduction

The professionalization of women's sports, especially football, considerably raised the demand for physical training in recent years [1,2], leading to an increase in injuries [3–5]. Epidemiological data indicate that women have a two to four times higher risk of anterior cruciate ligament (ACL) tears than men [6]. This percentage can be increased up to eight times in female athletes [7–9]. Since ACL injuries have a considerable impact on quality of life and medical costs, it is relevant to determine the injury risk caused by the difference in knee laxity according to the phases of the menstrual cycle. For this reason, knowledge of the correlation between physiological variation in women during the menstrual cycle and the risk of injury should be a key issue in the 21st century [10].

The menstrual cycle is defined as a period designating "physiological changes that occur in fertile women and other female primates for the purpose of sexual reproduction" [11]. In women, the menstrual cycle has an average duration of 28 days. During this period, hormonal fluctuations are present and make it possible to differentiate four phases: pre-ovulatory, ovulatory, post-ovulatory and menstrual. These episodes are controlled by several hormones, ovarian and steroidal, which each have different actions. During the pre-ovulatory or follicular phase, there is an increase in estrogen and follicle-stimulating hormone (FSH) levels [12]. Ovulation, on the other hand, is caused by a spike in luteinizing hormone (LH) in synergy with a spike in FSH, produced by an increase in estrogen. Successively, the post-ovulatory or luteal phase is characterized by an increase in the concentration of progesterone and inhibition of FSH and LH [13].

The menstrual cycle can be regulated by taking exogenous hormones, for medical reasons or as a contraceptive method. For these reasons, the use of oral contraceptives (OC) is increasingly common among young women. To date, no consensus could be reached on the positive or negative effect of OC on the incidence of ACL injuries [14].

Knee laxity is characterized by the anterior displacement of the tibia on the femur when a load is applied to the posterior surface of the tibia. The main stabilizer of the knee, limiting this movement, is the ACL [13]. During excessive anterior translations, this ligament is put under tension, and can therefore be subject to ruptures. Many studies have reported greater anterior laxity at the knee in women compared to men, which may lead to a greater risk of ACL injury [15–17], especially with the onset of puberty [18]. Therefore, hormonal fluctuations in women must be taken into consideration. Indeed, steroid hormones influence ligament metabolism, as demonstrated by Liu et al. on the ACL [19]. The estrogens and progesterone exert an action on the synthesis of collagen, modifying the fibroblastic property and therefore its mechanical resistance to tension. In fact, correlations between increased blood estrogen concentration and reduced collagen levels have been observed, due to a decrease in the number of fibroblasts in the ACL [20]. Studies show that a high level of estradiol is responsible for the decrease in fibroblast proliferation and therefore the synthesis of ligament collagen [19,21,22], which affects ligament resistance, increasing its elasticity and its vulnerability to injury. Looking at the composition of the ACL, Faryniarz et al. have identified receptors for relaxin, a hormone known for its elastic property in tissues [23]. These findings suggest an action of this hormone on ACL laxity, which is predominant in women [23].

Previously, two systematic reviews have been conducted to determine the association between ACL tears, knee laxity and menstrual cycle [6,24]. However, more recent data have been published since their research. Therefore, the objective of this systematic review is to update the scientific knowledge about the influence of the menstrual cycle and OC on knee laxity and ACL injuries.

2. Methods

2.1. Study Design

A systematic review was performed following the recommendations of Preferred Reporting Items for Systematic Reviews and Meta-Analyses (PRISMA 2020 criteria). This review was registered in PROSPERO with the code number CRD42022345257.

2.2. Data Source

The searches were performed by two independent investigators in the PubMed, Scopus and Web of Sciences databased. A search equation using MESH terms was defined to obtain a precise search: ("menstrual cycle" OR "oral contraceptive" OR "hormones") AND ("knee laxity" OR "Anterior Cruciate Ligament injuries" OR "Anterior Cruciate Ligament laxness").

2.3. Inclusion/Exclusion Criteria

A publication date filter was added from 2015 to July 2022 to update the scientific knowledge, since the studies of the last systematic review on this subject (6) dated from 2015. After duplicates were removed, studies were excluded if they did not match the following inclusion and exclusion criteria: (i) article written in English; (ii) population: women aged 13–49 years, having menarche for at least a year; (iii) not post-menopausal women, pregnant women, women with genital pathologies, or women with serious pathologies (cancer); (iv) observational studies only.

2.4. Data Extraction and Data Synthesis

In the selected studies, certain data were extracted, in particular the objective and the type of study, the number and age of the participants and their specificity (sports or not) (Table 1). In addition, outcome measures such as ACL injury, laxity/stiffness of the knee and general joint laxity (GJL) were selected (Table 2).

Table 1. Studies included in this review.

Authors	Year	Objective	Type of Study	Participants	Age	Specificity	Study Quality
DeFroda et al. [25]	2019	To examine the effects of OC usage on ACL tear and subsequent reconstruction	Case-controls	82,874	15–49 years	Database	E
Gray et al. [26]	2016	To determine if women undergoing anterior cruciate ligament surgical reconstruction were less likely to use oral contraceptives than a matched noninjured population	Case-controls	12,819	15–39 years	Database	E
Herzog et al. [27]	2020	Quantifying the association between OC use and ACL injury	Cohort	2,992,084	13–45 years	Database	A
Khowailed et al. [28]	2015	Investigating the effects of 17^{β}-estradiol across phases of the menstrual cycle on neuromuscular control patterns of ACL laxity during running	transverse	12	25.6 ± 3.7 years	Runners	E
Lee et al. [29]	2015	To study the difference in ACL laxity after squat exercise in healthy women between OC users and non-OC users.	Cohort	40	25.2 ± 2.2 years	Low to moderate PA	E
Maruyama et al. [12]	2021	To determine the relationship between knee joint looseness, stiffness, and general joint looseness in relation to the menstrual cycle	transverse	15	21 ± 0.2 years	female students Low to moderate PA	E
Nose-Ogura et al. [30]	2017	To study the concentration of relaxin-2 during the menstrual cycle in athletes without and with OC.	transverse	106	22 ± 3 years	athletes	E
Shafiei et al. [9]	2016	Comparing knee laxity changes in the menstrual cycle in female athletes	transverse	40	25.5 ± 5.12 years	athletes	A
Shagawa et al. [31]	2021	To examine changes in AKL, stiffness, GR and GJL during the late follicular phase and ovulation	transverse	15	21 ± 0.3 years	E	E

Table 1. *Cont.*

Authors	Year	Objective	Type of Study	Participants	Age	Specificity	Study Quality
Stijak et al. [32]	2015	To determine the difference in testosterone, 17^{β} estradiol and progesterone concentrations between patients with and without ACL tear, and the possible effect of these hormones on generalized joint laxity	Case-controls	24	16–37 years	athletes	E

Abbreviations: OC, oral contraceptives; ACL, anterior cruciate ligament; AKL, anterior displacement of the tibia relative to the femur; GR, genu recurvatum; GJL, general joint laxity.

Table 2. Summary of included studies.

Authors	Variables Studied	Method of Measurement of Variables	ACL Tear	Contraceptive Method	Hormones Studied	Results
DeFroda et al. 2019 [25]	ACL injury	Follow-up over time	Case: surgical reconstruction of ACL with OC Controls: surgical reconstruction of ACL without OC	569 ACL reconstruction patients in the non-OC group, and 465 patients in the OC group.	/	OC users: 18% decrease in the risk of ACL tear requiring reconstruction OC users 15–19 age group: 63% reduction in tears
Gray et al. 2016 [26]	ACL injury	Follow-up over time	Case: surgical reconstruction of the ACL	12,819 OC	/	15–19 age group with ACL repair surgery: not use OC 1.22 times > controls (12 months prior to injury) ($p < 0.0001$) 25–39 age group: use OC 1.1 to 1.16 times > controls OC users: 18% fewer ACL injuries than non-users (15–19 age group)
Herzog et al. 2020 [27]	ACL injury	Follow-up over time	3571 at OC 1620 at IUD	OC (exposed) IUD (unexposed)	/	No difference in the risk of ACL injury in OCs and IUDs
Khowailed et al. 2015 [28]	Knee laxity [estradiol] Neuro-muscular control	KT-2000 Cobas e602 blood test (Roche/ Hitachi) → EMG	No	No	Estradiol	↑ ATT: ovulation (5.75 ± 0.47) ($p < 0.01$) ↓ ATT: follicular (4.18 ± 0.27) ($p < 0.01$) [17^{β} estradiol] ↓ menstruation (34.14 ± 15.47 pg/mL) [17^{β} estradiol] ↑ ovulation (207.74 ± 53.42 pg/mL) $[17^{\beta}\text{ estradiol}]_{ovulation} > [17^{\beta}\text{ estradiol}]_{follicular}$ ($p < 0.01$)
Lee et al. 2015 [29]	ACL laxity Pain	KT-2000 EVA	No	25 with normal cycle 15 with OC	/	Decreased ACL laxity (OC users) compared to non-OC users before and after strenuous exercise Pain after exercise (OC) > pain after exercise (non-OC)

Table 2. *Cont.*

Authors	Variables Studied	Method of Measurement of Variables	ACL Tear	Contraceptive Method	Hormones Studied	Results
Maruyama et al. 2021 [12]	AKL Stiffness GJL GR	KS-measurement Δ force/Δ displacement University of Tokyo joint laxity test Item 4 of the GJL test, goniometer	No	No	/	GR: AKL (89N and 133N) Ovulatory phase > early follicular phase No GR: No difference in AKL between phases AKL: GR (ovulation) > no GR GJL: No differences between GR and non-GR AKL, stiffness, GJL: No difference between phases
Nose-Ogura et al. 2017 [30]	[relaxin-2] [sex hormones]	Immunoassay (quantikin Human relaxin-2) CLIA kit	No	16 with OC 77 eumenorhea 13 amenorrhea	relaxin-2 sex hormones	$[\text{relaxin-2}]_{\text{luteal}}$ > $[\text{relaxin-2}]_{\text{follicular, ovulatory}}$ in eumhenorrea [relaxin-2] at eumennorhee > [relaxin-2] at OC luteal phase: [relaxin-2] > 6.0 pg/mL in 36.8% of FDR cases. ACL injury correlation between relaxin-2 and progesterone in CM phases except luteal phase
Shafiei et al. 2016 [9]	Knee laxity (ACL) [sex hormones]	Lachman Front drawer ELISA and DEMEDITEC kit	No	No	[estrogen] [progesterone]	No difference for ACL laxity nor hormonal levels. No relationship between menstrual cycle phases and knee laxity
Shagawa et al. 2021 [31]	AKL stiffness GJL GR [estradiol]	KS-measurement Δ force/Δ displacement University of Tokyo joint laxity test Item 4 of the GJL test, goniometer Saliva sample Ovulation kit (+T° control: thermometer)	No	No	Estradiol	$[\text{oestradiol}]_{\text{ovulatory}}$ > $_{\text{late follicular [oestradiol]}}$ ($p = 0.018$) No differences between AKL or rigidity of the follicular phase and the ovulatory phase GR and GJL ↑ during ovulatory phase > follicular phase Changes in [estradiol] during the cycle can affect changes in GR and GJL
Stijak et al. 2015 [32]	GJL [sex hormones]	Laxity score (Beighton, Solomon, Soskolne) sample → enzyme immunoassay Salimetrics	Yes contactless (case) No (control)	No	[17^{β} estradiol] [progesterone] [testosterone]	Significant differences in [17^{β} estradiol], [progesterone], and [testosterone] between the two groups Lower [hormones] in women with ACL tears

Abbreviations: OC, oral contraceptives; ACL, anterior cruciate ligament; IUD, intrauterine device; AKL, anterior displacement of the tibia relative to the femur; GR, genu recurvatum; GJL, general joint laxity; ↓ significant decrease; ↑ significant increase.

2.5. Risk of Bias Assessment

The assessment of article quality and risk of bias was performed using McMaster guidelines for Critical Review Form for Quantitative studies [33]. This guide is made up of 16 criteria: (1) objectives of the study clearly defined; (2) presentation of the relevant context; (3) design of the study appropriate; (4) sample described in detail; (5) sample size justified; (6) reliable measurement of variables; (7) valid measurement of variables; (8)

intervention described in detail; (9) contamination avoided; (10) co-intervention avoided; (11) intervention replicable; (12) results reported in terms of statistical significance; (13) methods analysis; (14) reported clinical importance; (15) loss and abandons reported; (16) appropriate conclusions regarding study methods and results. In this review, consisting exclusively of observational studies, criteria 9 and 10 (contaminations and co-interventions avoided) could not be evaluated. For each item, 1 point was given if it was completed and 0 when it was not answered. The quality score of each study was calculated on 14 criteria, 14 being the maximum score. Then, a classification according to the methodological quality was carried out in percentage form. When the total score was lower than 50% the quality was considered low, acceptable between 50% and 64%, high quality between 65% and 79%, and excellent when the result was above 80%.

3. Results

A total of ten articles was selected in this systematic review according to the inclusion and exclusion criteria (Tables 1 and 2). The steps in this data collection are represented in Figure 1. Among these ten studies, the sports levels of the women differed: four studies included athletes [9,28,30,32]; three others involved women doing low-to-moderate-intensity physical activity [12,29,31]; and the three remaining studies did not provide this information, as they were extracted from a database [25–27].

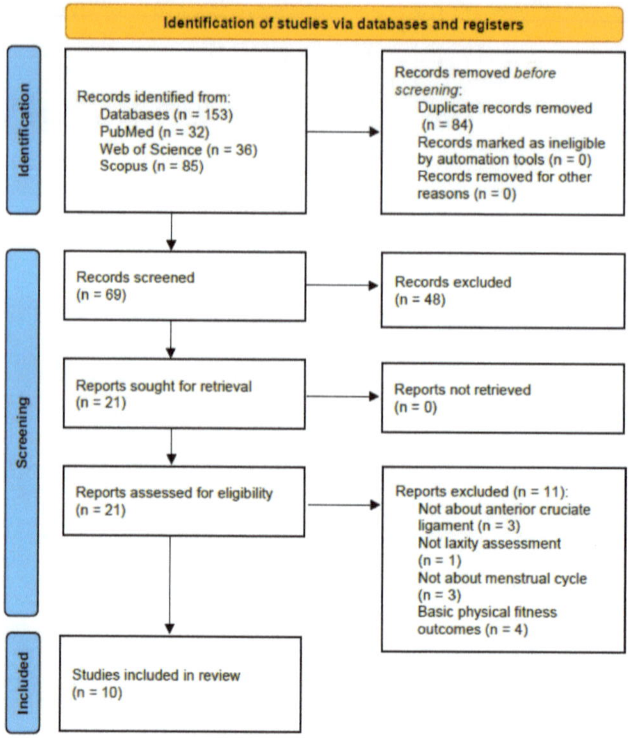

Figure 1. PRISMA Flow diagram.

The sample size ranged from 12 to 2,992,084 women aged 13–49 years old. The studies included in this review had scores based on the McMaster guidelines for Critical Review Form for Quantitative studies between 8/14 and 13/14. Three studies (30%) showed acceptable methodological quality, one (10%) high quality, and the other six (60%) were rated as excellent (Table 3).

Table 3. Quality Results methodology of the studies included in this review.

	Criteria															T	%	QM
Article	**1**	**2**	**3**	**4**	**5**	**6**	**7**	**8**	**9**	**10**	**11**	**12**	**13**	**14**	**15**			
Shagawa et al., (2021) [31]	1	1	1 (c)	1	0	1	1	1	N/A	1	1	1	1	1	1	13/14	93%	E
Maruyama et al., (2021) [12]	1	1	1 (c)	1	0	1	1	1	N/A	1	1	1	1	1	1	13/14	93%	E
Herzog et al., (2020) [27]	1	1	1 (a)	0	0	1	1	0	N/A	0	0	1	1	0	1	8/14	57%	A
DeFroda et al., (2019) [25]	1	1	1 (b)	0	0	1	1	0	N/A	0	1	1	1	0	1	9/14	64%	A
Nose-Ogura et al., (2017) [30]	1	1	1 (c)	1	0	1	1	1	N/A	1	1	1	1	1	1	13/14	93%	E
Gray et al., (2016) [26]	1	1	1 (b)	1	1	1	1	0	N/A	0	1	1	1	0	1	11/14	79%	H
Shafiei et al., (2016) [9]	1	1	1 (c)	1	0	1	1	0	N/A	0	1	1	0	0	0	8/14	57%	A
Lee et al., (2015) [29]	1	1	1 (a)	1	0	1	1	1	N/A	1	1	1	1	0	1	12/14	86%	E
Stijak et al., (2015) [32]	1	1	1 (b)	1	0	1	1	1	N/A	1	1	1	1	0	1	12/14	86%	E
Khowailed et al., (2015) [28]	1	1	1 (c)	1	1	1	1	1	N/A	1	1	1	1	0	1	13/14	93%	E
Total	8	8	8	7	1	8	8	6	/	6	7	8	7	3	7			

Abbreviations: 1: Objective of the study clearly defined; 2: Presentation of the relevant context; 3: Appropriate study design; 4: Sample described in detail; 5: Sample size justified; 6: Measurements of reliable variables; 7: Measurements of valid variables; 8: Intervention described in detail; 9: Contaminations and co-interventions avoided; 10: Replicable intervention; 11: Results reported in terms of statistical significance; 12: Appropriate analytical methods; 13: Reported clinical importance; 14: Reported Losses and Abandonments; 15: Appropriate conclusion regarding the methods and results of the study; T: total; %: Percentage of evaluation criteria completed; QM: Methodological quality; N/A: Not applied; E: Excellent; H: High quality; A: Acceptable; (a): cohort; (b): control cases; (c): transverse.

3.1. ACL Tears

Herzog et al., using a database, found no correlation between ACL injury risk and OC uses compared to an intrauterine device [27]. At the same time, the study conducted by Gray et al. demonstrated an 18% decrease in ACL reconstructions in OC users compared to non-users in the 15–19-year age group, with the mean age of cases being 24.11 years. In this group, women who had undergone ACL-repair surgery were 1.22 times more likely not to use OC than controls in the 12 months before the injury ($p < 0.001$) [26]. In this same guideline, De Froda et al. demonstrated a 63% reduction in ACL tear rate in OC users in the same age group (15–19-years) [25].

In the study carried out by Stijak et al., ACL tears were caused by an indirect mechanism (without contact) [34]. There were no significant differences in the joint laxity score between the participants with the ACL intact and the participants suffering an ACL tear [32].

3.2. Knee Laxity

Modulation of knee laxity during the menstrual cycle has been reported in five studies [9,12,28,29,31]. Laxity, illustrated as the anterior displacement of the tibia (AKL) relative to the femur, was assessed by different methods: an instrumental technique (KT2000, KS measurement), and orthopedics test (Lachman's test, and the University of Tokyo joint-laxity test). Khowailed et al. reported a statistically significant increase ($p < 0.01$) in tibial anterior translation (TAT) during the ovulatory phase compared to the follicular phase [28]. However, among non-genu-recurvatum patients (GR), no significant differences were observed between the phases of the menstrual cycle [12]. Similarly, Shagawa et al. revealed an increased generalized joint laxity (GJL) during the ovulatory phase compared to the follicular phase, but no differences were found between the two phases concerning stiffness or AKL [31]. Finally, Maruyama et al. did not observe significant differences in GJL between GR and non-GR [12]. Among users of OC, researchers noted a decreased ACL laxity [29].

3.3. Hormones

Hormonal fluctuations were measured in five articles using kit ovulatory and blood and saliva samples [9,28,30–32]. Two studies showed that the estradiol concentration during ovulation was significantly higher than during the follicular phase [28,31]. In addition, the concentrations of estradiol, progesterone and testosterone were significantly reduced in women with a torn ACL [32]. However, Shafei et al. did not observe significant differences between ACL laxity and hormonal levels [9]. Finally, during the luteal phase, the concentration of relaxin was greater than 6.0 pg/mL in 36.8% of cases, proving to be a risk factor for rupture of the ACL [30].

4. Discussion

This study provides novel research about the influence of the menstrual cycle and OC on knee laxity and ACL injuries. The menstrual cycle seems to influence knee laxity in women, the latter being higher during the ovulatory phase. However, current knowledge cannot yield conclusions about the influence of the menstrual cycle on the risk of ACL injury. OC may decrease the possibility of an ACL tear.

As demonstrated by Stijak et al., women with an ACL tear show a decrease in the concentration of testosterone, 17B estradiol and progesterone. This hormonal drop could lead to ACL damage and could be considered a risk factor. This, therefore, raises questions about the protective action or lack thereof of hormones. Indeed, during the ovulatory phase, when estradiol concentration is maximal, joint stability would be negatively affected [35]. This estradiol peak during the ovulatory phase is confirmed in the study carried out by Khowailed et al. since the concentration of 17B estradiol significantly increases during ovulation, compared to the follicular phase [28]. These findings are in accordance with the results of the study performed by Shagawa et al. [31].

Regarding knee laxity, an increase in TAT could be observed during the ovulatory phase compared to the follicular phase [28]. In this line, Shagawa et al. observed an increase in GJL during the menstrual cycle phase where estradiol is higher [31]. Therefore, it can be assumed that the increase in estrogen concentration would induce an increase in knee laxity characterized by TAT. In concordance, the review performed by Herzberg et al. in 2017 identified a higher ACL laxity during the ovulatory phase compared to the follicular phase [24].

The role of OC in regulating the menstrual cycle has been investigated in some studies [27,29,30]. Herzog et al. did not detect an increased risk of ACL injury in OC or intrauterine device (IUD) users, not allowing a protective effect to be established. Nevertheless, the control group (IUD) had exogenous hormones, so it would be interesting to study the difference with eumenorrheic women (women with a regular cycle and not taking OC). According to Nose Ogura et al. and Lee et al., OC users have decreased ACL laxity [29] due to a lower relaxin concentration [30], and therefore may have a lower risk of an ACL tear.

Consistent with these findings, the study carried out by Gray et al. suggests that the regulatory effects of OC on estrogen levels may reduce the overall risk of ACL injury in young women by reducing the risk factor for induced ACL laxity by estrogen [26]. Despite the limited clinical evidence on this protective aspect, De Froda et al. state that it would be advisable to consider the use of OC in elite high school and college-aged athletes, especially those who are at risk of ACL tear [25]. Future research is needed to confirm this protective association.

Some aspects should be considered when interpreting the results of this review. The age range of the participants is very wide, between 13 and 49 years old. In addition, the selected studies do not specify whether the participants had been pregnant before. During pregnancy, women suffer an important hormonal imbalance [36], so it would be wise to carry out studies on women who have never been pregnant to observe the true impact of a menstrual cycle without previous upheaval. It can be assumed that after pregnancy the hormonal levels are modified, in particular those of relaxin, so it would be interesting to observe whether there is an increased risk of ACL injury in postpartum women.

Another important aspect to consider is that the hormonal profile of each woman may differ during each cycle even in women with a regular menstrual cycle [37]. For this reason, measurement tools should be predefined identically in each study to avoid any bias in the evaluation of data, allowing researchers to perform a meta-analysis on this subject. The human body being very complex, it would be necessary to study the interaction of the various hormones and not their individual action. Stijak et al. have shown that a higher concentration of testosterone could act protectively for ACL injuries in women [32]. Interestingly, Lovering and Romani [10] demonstrated the presence of androgen receptors in the female ACL. In fact, testosterone, which plays a role in connective tissue remodeling, is influenced by the presence of other hormones. In fact, an increase in the sex-hormone-binding globulin (SHBG) levels is associated with decreased levels of testosterone. However, estrogen and testosterone have an antagonistic relationship. Higher rates of free testosterone have been correlated with an increase in ACL stiffness during the ovulatory phase [10]. Further research on this topic is needed.

Studies have observed changes in the menstrual cycle associated with physical activity levels [38]. Indeed, most women (>50%) practicing regular physical activity have menstrual dysfunction, characterized by an irregular menstrual cycle [36]. It would therefore be interesting to study the differences in knee laxity related to the menstrual cycle between physically active women and sedentary women.

Another factor that could exert a negative influence on the menstrual cycle is stress. Moreover, progesterone levels, which are higher during the luteal phase, may be associated with anxiety [39]. An increase in mental fatigue has also been observed during this phase [40], which could contribute to certain injuries, in particular the ACL, according to many studies [38].

The previous review published in 2015 by Somerson et al. revealed a greater anterior translation of the knee in the ovulatory and luteal phases of the menstrual cycle. In this current review, we find the same results, which seem consistent. However, no correlation was found between the risk of ACL rupture and periods of laxity.

Further studies are needed to observe the association between estrogen levels and knee joint laxity. Future research should be considered to observe the action of testosterone in women especially in sports. A better understanding of women's physiology is essential to injury prevention.

5. Conclusions

In conclusion, the menstrual cycle, due to hormonal fluctuation, seems to influence knee laxity in women. Female athletes have a higher risk of ACL rupture due to hormonal changes, especially relaxin. The increase in the latter was observed predominantly in the luteal phase parallel to estradiol during the ovulatory phase when the TAT was greater at the knee. On the other hand, OC decreased the laxity of the ACL, making the TAT weaker and decreasing the levels of relaxin, which appears to decrease the possibility of an ACL tear. However, based on the literature, we cannot conclude that there is a correlation between the menstrual cycle and the risk of ACL injury.

Author Contributions: J.M. and A.F., Conceptualization, data curation, formal analysis, writing-original draft; D.D.-B. and Á.G.-d.-l.-F., investigation, methodology, validation, writing—review & editing; J.S.-A., Conceptualization, methodology, supervision, validation; G.G.-P.-d.-S., methodology, supervision, validation, writing—review & editing. All authors have read and agreed to the published version of the manuscript.

Funding: The authors received no financial support for the research, authorship, and/or publication of this article.

Institutional Review Board Statement: Not applicable.

Informed Consent Statement: Not applicable.

Data Availability Statement: Not applicable.

Conflicts of Interest: The authors declare no conflict of interest.

References

1. Datson, N.; Drust, B.; Weston, M.; Jarman, I.H.; Lisboa, P.; Gregson, W. Match Physical Performance of Elite Female Soccer Players During International Competition. *J. Strength Cond. Res.* **2017**, *31*, 2379–2387. [CrossRef] [PubMed]
2. Datson, N.; Hulton, A.; Andersson, H.; Lewis, T.; Weston, M.; Drust, B.; Gregson, W. Applied physiology of female soccer: An update. *Sport. Med.* **2014**, *44*, 1225–1240. [CrossRef] [PubMed]
3. Gebert, A.; Gerber, M.; Pühse, U.; Gassmann, P.; Stamm, H.; Lamprecht, M. Costs resulting from nonprofessional soccer injuries in Switzerland: A detailed analysis. *J. Sport Health Sci.* **2020**, *9*, 240–247. [CrossRef]
4. Mufty, S.; Bollars, P.; Vanlommel, L.; Van Crombrugge, K.; Corten, K.; Bellemans, J. Injuries in male versus female soccer players: Epidemiology of a nationwide study. *Acta Orthop. Belg.* **2015**, *81*, 289–295. [PubMed]
5. Roos, K.G.; Wasserman, E.B.; Dalton, S.L.; Gray, A.; Djoko, A.; Dompier, T.P.; Kerr, Z.Y. Epidemiology of 3825 injuries sustained in six seasons of National Collegiate Athletic Association men's and women's soccer (2009/2010–2014/2015). *Br. J. Sports Med.* **2017**, *51*, 1029–1034. [CrossRef] [PubMed]
6. Somerson, J.S.; Isby, I.J.; Hagen, M.S.; Kweon, C.Y.; Gee, A.O. The Menstrual Cycle May Affect Anterior Knee Laxity and the Rate of Anterior Cruciate Ligament Rupture: A Systematic Review and Meta-Analysis. *JBJS Rev.* **2019**, *7*, e2. [CrossRef]
7. Larruskain, J.; Lekue, J.A.; Diaz, N.; Odriozola, A.; Gil, S.M. A comparison of injuries in elite male and female football players: A five-season prospective study. *Scand. J. Med. Sci. Sports* **2018**, *28*, 237–245. [CrossRef]
8. Lin, C.Y.; Casey, E.; Herman, D.C.; Katz, N.; Tenforde, A.S. Sex Differences in Common Sports Injuries. *PM&R* **2018**, *10*, 1073–1082.
9. Shafiei, S.E.; Peyvandi, S.; Kariminasab, M.H.; Azar, M.S.; Daneshpoor, S.M.M.; Khalilian, A.; Aghajantabar, Z. Knee Laxity Variations in the Menstrual Cycle in Female Athletes Referred to the Orthopedic Clinic. *Asian J. Sports Med.* **2016**, *7*, e30199. [CrossRef]
10. Lovering, R.M.; Romani, W.A. Effect of testosterone on the female anterior cruciate ligament. *Am. J. Physiol. Integr. Comp. Physiol.* **2005**, *289*, R15–R22. [CrossRef]
11. Gosselin, M. *Menstrual Cycle: Signs and Symptoms, Psychological/Behavioral Changes and Abnormalities: Signs and Symptoms, Psychological/Behavioral Changes and Abnormalities*; Nova Science Publishers Incorporated: Hauppauge, NY, USA, 2013.

12. Maruyama, S.; Yamazaki, T.; Sato, Y.; Suzuki, Y.; Shimizu, S.; Ikezu, M.; Kaneko, F.; Matsuzawa, K.; Hirabayashi, R.; Edama, M. Relationship Between Anterior Knee Laxity and General Joint Laxity During the Menstrual Cycle. *Orthop. J. Sports Med.* **2021**, *9*, 2325967121993045. [CrossRef] [PubMed]
13. Knudtson, J.; McLaughlin, J.E. *Endocrinologie de la Reproduction Féminine*; MSD Manuals: Rahway, NJ, USA, 2019.
14. Samuelson, K.; Balk, E.M.; Sevetson, E.L.; Fleming, B.C. Limited Evidence Suggests a Protective Association Between Oral Contraceptive Pill Use and Anterior Cruciate Ligament Injuries in Females: A Systematic Review. *Sports Health A Multidiscip. Approach* **2017**, *9*, 498–510. [CrossRef] [PubMed]
15. Nguyen, A.-D.; Shultz, S.J. Sex Differences in Clinical Measures of Lower Extremity Alignment. *J. Orthop. Sports Phys. Ther.* **2007**, *37*, 389–398. [CrossRef]
16. Shultz, S.J.; Sander, T.C.; Kirk, S.E.; Perrin, D.H. Sex differences in knee joint laxity change across the female menstrual cycle. *J. Sports Med. Phys. Fitness* **2005**, *45*, 594–603.
17. Beynnon, B.D.; Bernstein, I.M.; Belisle, A.; Brattbakk, B.; Devanny, P.; Risinger, R.; Durant, D. The Effect of Estradiol and Progesterone on Knee and Ankle Joint Laxity. *Am. J. Sports Med.* **2005**, *33*, 1298–1304. [CrossRef] [PubMed]
18. Shultz, S.J.; Nguyen, A.-D.; Schutz, R.J. Differences in Lower Extremity Anatomical and Postural Characteristics in Males and Females Between Maturation Groups. *J. Orthop. Sports Phys. Ther.* **2008**, *38*, 137–149. [CrossRef] [PubMed]
19. Liu, S.H.; Al-Shaikh, R.; Panossian, V.; Yang, R.S.; Nelson, S.D.; Soleiman, N.; Finerman, G.A.M.; Lane, J.M. Primary immunolocalization of estrogen and progesterone target cells in the human anterior cruciate ligament. *J. Orthop. Res.* **1996**, *14*, 526–533. [CrossRef]
20. Yu, W.D.; Panossian, V.; Hatch, J.D.; Liu, S.H.; A Finerman, G. Combined Effects of Estrogen and Progesterone on the Anterior Cruciate Ligament. *Clin. Orthop. Relat. Res.* **2001**, *383*, 268–281. [CrossRef]
21. Karageanes, S.J.; Blackburn, K.; Vangelos, Z.A. The association of the menstrual cycle with the laxity of the anterior cruciate ligament in adolescent female athletes. *Clin. J. Sport Med.* **2000**, *10*, 162–168. [CrossRef]
22. Schmitz, R.J.; Shultz, S.J. Anterior knee stiffness changes in laxity "responders" versus "nonresponders" across the menstrual cycle. *J. Athl. Train.* **2013**, *48*, 39–46. [CrossRef]
23. Faryniarz, D.A.; Bhargava, M.; Lajam, C.; Attia, E.T.; Hannafin, J.A. Quantitation of estrogen receptors and relaxin binding in human anterior cruciate ligament fibroblasts. *Vitr. Cell. Dev. Biol. Anim.* **2006**, *42*, 176–181. [CrossRef] [PubMed]
24. Herzberg, S.D.; Motu'apuaka, M.L.; Lambert, W.; Fu, R.; Brady, J.; Guise, J.-M. The Effect of Menstrual Cycle and Contraceptives on ACL Injuries and Laxity: A Systematic Review and Meta-analysis. *Orthop. J. Sport. Med.* **2017**, *5*, 2325967117718781. [CrossRef] [PubMed]
25. DeFroda, S.F.; Bokshan, S.L.; Worobey, S.; Ready, L.; Daniels, A.H.; Owens, B.D. Oral contraceptives provide protection against anterior cruciate ligament tears: A national database study of 165,748 female patients. *Phys. Sportsmed.* **2019**, *47*, 416–420. [CrossRef] [PubMed]
26. Gray, A.M.; Gugala, Z.; Baillargeon, J.G. Effects of Oral Contraceptive Use on Anterior Cruciate Ligament Injury Epidemiology. *Med. Sci. Sports Exerc.* **2016**, *48*, 648–654. [CrossRef] [PubMed]
27. Herzog, M.M.; Young, J.C.; Lund, J.L.; Pate, V.; Mack, C.D.; Marshall, S.W. Oral contraceptive use and anterior cruciate ligament injury: Comparison of active comparator new user cohort and case-control study designs. *Inj. Epidemiol.* **2020**, *7*, 53. [CrossRef]
28. Khowailed, I.A.; Petrofsky, J.; Lohman, E.; Daher, N.; Mohamed, O. 17$^{\beta}$-Estradiol Induced Effects on Anterior Cruciate Ligament Laxness and Neuromuscular Activation Patterns in Female Runners. *J. Womens Health* **2015**, *24*, 670–680. [CrossRef]
29. Lee, H.; Petrofsky, J.S.; Yim, J. Do Oral Contraceptives Alter Knee Ligament Damage with Heavy Exercise? *Tohoku J. Exp. Med.* **2015**, *237*, 51–56. [CrossRef]
30. Nose-Ogura, S.; Yoshino, O.; Yamada-Nomoto, K.; Nakamura, M.; Harada, M.; Dohi, M.; Okuwaki, T.; Osuga, Y.; Kawahara, T.; Saito, S. Oral contraceptive therapy reduces serum relaxin-2 in elite female athletes. *J. Obstet. Gynaecol. Res.* **2017**, *43*, 530–535. [CrossRef]
31. Shagawa, M.; Maruyama, S.; Sekine, C.; Yokota, H.; Hirabayashi, R.; Hirata, A.; Yokoyama, M.; Edama, M. Comparison of anterior knee laxity, stiffness, genu recurvatum, and general joint laxity in the late follicular phase and the ovulatory phase of the menstrual cycle. *BMC Musculoskelet. Disord.* **2021**, *22*, 886. [CrossRef]
32. Stijak, L.; Kadija, M.; Djulejić, V.; Aksić, M.; Petronijević, N.; Marković, B.; Radonjić, V.; Bumbaširević, M.; Filipović, B. The influence of sex hormones on anterior cruciate ligament rupture: Female study. *Knee Surg. Sport. Traumatol. Arthrosc.* **2015**, *23*, 2742–2749. [CrossRef]
33. Law, M.; Stewart, D.; Pollock, N.; Letts, L.; Bosch, J.; Westmorland, M. *Guidelines for Critical Review Form—Quantitative Studies*; McMaster University Occupational Therapy Evidence-Based Practice Research Group: Hamilton, ON, Canada, 1998.
34. Bisciotti, G.N.; Chamari, K.; Cena, E.; Bisciotti, A.; Corsini, A.; Volpi, P. Anterior cruciate ligament injury risk factors in football. *J. Sports Med. Phys. Fitness* **2019**, *59*, 1724–1738. [CrossRef] [PubMed]
35. Sung, E.-S.; Kim, J.-H. The influence of ovulation on postural stability (Biodex Balance System) in young female. *J. Exerc. Rehabil.* **2018**, *14*, 638–642. [CrossRef] [PubMed]
36. Noyola-Martínez, N.; Halhali, A.; Barrera, D. Steroid hormones and pregnancy. *Gynecol. Endocrinol.* **2019**, *35*, 376–384. [CrossRef] [PubMed]
37. Hackney, A. *Sex Hormones, Exercise and Women*, 1st ed.; Springer: Cham, Switzerland, 2017; pp. 113–138.

38. Orio, F.; Muscogiuri, G.; Ascione, A.; Marciano, F.; Volpe, A.; La Sala, G.; Savastano, S.; Colao, A.; Palomba, S. Effects of physical exercise on the female reproductive system. *Minerva Endocrinol.* **2013**, *38*, 305–319.
39. Reynolds, T.A.; Makhanova, A.; Marcinkowska, U.M.; Jasienska, G.; McNulty, J.K.; Eckel, L.A.; Nikonova, L.; Maner, J.K. Progesterone and women's anxiety across the menstrual cycle. *Horm. Behav.* **2018**, *102*, 34–40. [CrossRef]
40. Nillni, Y.I.; Toufexis, D.J.; Rohan, K.J. Anxiety sensitivity, the menstrual cycle, and panic disorder: A putative neuroendocrine and psychological interaction. *Clin. Psychol. Rev.* **2011**, *31*, 1183–1191. [CrossRef]

applied sciences

MDPI

Article

Toward Accessible Hearing Care: The Development of a Versatile Arabic Word-in-Noise Screening Tool: A Pilot Study

Ahmad O. Alokaily , Abdulaziz F. Alqabbani, Adham Aleid and Khalid Alhussaini *

Department of Biomedical Technology, College of Applied Medical Sciences, King Saud University, Riyadh 12372, Saudi Arabia
* Correspondence: kalhussaini@ksu.edu.sa

Abstract: Speech-in-noise tests are used to assess the ability of the human auditory system to perceive speech in a noisy environment. Early diagnosis of hearing deficits helps health professionals to plan for the most appropriate management. However, hospitals and auditory clinics have a shortage of reliable Arabic versions of speech-in-noise tests. Additionally, access to specialized healthcare facilities is associated with socioeconomic status. Hence, individuals with compromised socioeconomic status do not have proper access to healthcare. Thus, In the current study, a mobile and cost-effective Arabic speech-in-noise test was developed and tested on 30 normal-hearing subjects, and their ability to perceive words-in-noise was evaluated. Moreover, a comparison between two different background noises was explored (multi-talker babble noise and white noise). The results revealed a significant difference in the thresholds between the two types of background noises. The percent-correct scores ranged from 100% to 54.17% for the white background noise and 91.57% to 50% for the multi-talker babble background noise. The proposed Arabic word-in-noise screening tool has the potential to be used effectively to screen for deteriorated speech perception abilities, particularly in low-resource settings.

Keywords: word in noise; speech in noise; hearing; healthcare accessibility; rural; out of clinic; hearing impairment; diagnostic tool; cost-effective; Arabic

check for updates

Citation: Alokaily, A.O.; Alqabbani, A.F.; Aleid, A.; Alhussaini, K. Toward Accessible Hearing Care: The Development of a Versatile Arabic Word-in-Noise Screening Tool: A Pilot Study. *Appl. Sci.* **2022**, *12*, 12459. https://doi.org/10.3390/app122312459

Academic Editors: Alice Bruscolini, Massimo Ralli and Gaspare Palaia

Received: 1 November 2022
Accepted: 5 December 2022
Published: 6 December 2022

1. Introduction

With over 430 million cases worldwide, hearing loss is one of the most critical global health issues that must be addressed [1]. In addition, approximately 344 million sufferers of hearing loss (80% of total cases) reside in low- and middle-income countries, as reported by the World Health Organization [1]. Societies with low socioeconomic status and poor education often do not have the same access to specialized healthcare services as those with quality education and higher incomes [2], which in turn could significantly delay the diagnosis and treatment of hearing impairment. Moreover, hearing impairments have an indirect impact on the individual's social, occupational, educational, and psychological well-being [3–5]. For instance, people with hearing problems are highly susceptible to depression compared to people with normal hearing [6]. Despite the high prevalence of hearing impairments, deployment of diagnostic and therapeutic hearing healthcare is sparse [3]. It has been reported that the time gap between the initial onset of hearing impairment and the actual seeking for rehabilitation interventions for people with hearing problems is long, typically 10 years [7]. Thus, there is an emerging need to develop and validate accurate and highly sensitive hearing screening instruments and tests that are accessible with high fidelity, to help ensure the detection of hearing impairment at early stages to facilitate timely intervention.

Numerous hearing tests are currently deployed in hearing clinics to evaluate the integrity of the auditory system, assess hearing ability, and detect the degree of existing hearing loss. The pure-tone test, for example, is the most commonly used test in auditory clinics that determines the person's hearing ability by listening to tones at different frequencies. Another common clinical hearing test is the auditory brainstem responses test, in which the integrity of the neural pathway is investigated by measuring the electrophysiological responses to an acoustic stimulus delivered either monaurally or binaurally [8,9]. However, administering the most widely available hearing tests often requires costly audiometric booths to reduce the background noise [10]. Additionally, both of the aforementioned hearing tests lack important diagnostic features, such as the capability to identify the person's ability to comprehend complex stimuli as speech perception, which is a critical aspect of the patient's everyday life situations [11].

Perceiving speech with background noise can be challenging, even for people with normal hearing. Speech perception is not only the recognition of sound segments but also the ability of the listener to detect and interpret signals. Therefore, speech perception is defined as the act of labeling acoustic signals with appropriate linguistic symbols [12].

Several conditions are associated with speech perception difficulties, such as central auditory processing disorder (CAPD) [13]. Patients with CAPD face difficulties in comprehending spoken language in an environment with competing background noise. Children with CAPD frequently ask for repetitions, frequently saying "what?" or "huh?" [13]. Moreover, individuals with dyslexia may encounter difficulties in perceiving speech in noisy environments [14]. Furthermore, autism spectrum disorder is associated with speech perception difficulties and auditory processing deficits [15].

Speech perception tests are commonly used to screen for such disorders in various languages. Compared to standard hearing tests, speech-in-noise tests have advantages such as an inexpensive setup and the ability to efficiently administer the test outside the sound booth [16]. Clinically validated speech-in-noise tests vary based on the linguistics of the stimulus of interest, the specific type of background noise in which the speech is embedded, scoring methods, and the signal-to-noise ratio (SNR) presentation level [17].

One of the most common tests is the quick-speech-in-noise test (QuickSIN) [18]. The QuickSIN test estimates speech-in-noise performance by presenting short sentences with the presence of four-talker babbling in the background [18]. The test contains 12 lists that have six sentences in each list. The babble level increases in 5 dB increments while the sentences are presented at a fixed level. Another standard test is the Bamford Kowal-Bench speech-in-noise test (BKB-SIN). It contains 18 lists of sentences in the presence of a four-talker babbling noise in the background [19].

The Hearing in Noise Test for Children (HINT-C) is also used to diagnose speech perception conditions in children [20,21]. Finally, one of the most widely used tests in audiology clinics is the word-in-noise (WIN) test. The WIN test has the advantage of being relatively easier to administer and requires less working memory and linguistic context during the performance compared to the speech-in-noise test [22].

One of the few Arabic versions of the speech-in-noise tests is the Arabic matrix test. The Arabic matrix test generates sentences randomly from a matrix that are presented with and without background noise [23]. The Speech Perception of Words in Noise test is an Arabic WIN test in which words are presented at zero SNR in a cafeteria background noise [24]. Another example of an Arabic WIN test is the Pediatric Arabic Auditory Speech Test (PAAST), which was developed to investigate speech perception in children [25]. The main difference between the PAAST and the aforementioned Speech Perception of Words in Noise test is how the responses are collected. In the PAAST, children are asked to choose an image that corresponds to a presented word they heard [25].

Although speech-in-noise tests can provide a good assessment of speech perception inside hospitals and clinics, it is difficult to access specialized healthcare facilities for some individuals, communities, and countries, especially during pandemic and quarantine situations or even in rural areas. For that, several investigators have designed, developed, and validated computer-based applications for speech perception assessment outside the clinic to help with the early detection of hearing impairments. Outside clinic tests have the potential to alert individuals of possible hearing problems and a referral to secondary-level healthcare could confirm the level and type of the hearing impairment. For instance, computer-based or smartphone applications for audiological screening have shown reasonable efficiency [25–31].

For example, Hussein et al. [28] investigated the feasibility of implementing an existing mobile health application in a community-based setup to screen the hearing of young children. The study examined a mobile application, hearScreen^{TM}, which can be easily trained to be used by non-professional personnel such as school teachers, social workers, or nurses. The employment of the hearScreen^{TM} solution led to a referral rate of 24.9% in preschool children in a poor community. This study concluded that smartphone-based hearing screening could be a viable tool to detect unidentified hearing impairments with minimal training of non-professional individuals.

In addition, Bauer et al. [32] developed and validated an application named Ouviu that can be used pre-clinically to identify undetected hearing problems outside the clinical sound booth. The application conducts a hearing test that screens for 500, 1000, 2000, 5000, and 8000 Hz at low, medium, and high intensities. Results showed that Ouviu is capable of measuring environmental noise and outperformed the HearCheck screener equipment.

Alhussaini et al. [33] assessed auditory temporal resolution by utilizing an auditory gap-in-noise test. The authors compared the assessment of auditory temporal resolution in a controlled environment (inside a soundproof booth) versus assessing auditory temporal resolution in a normal room with an acceptable surrounding noise level (below 50 dB). The results showed that gap detection thresholds obtained outside the soundproof booth are reliable and comparable to the standard method currently used in a controlled clinical setting (inside a soundproof booth).

Furthermore, Govender and Mars [34] explored the outcomes of implementing asynchronous automated telehealth-based hearing screening and automated audiometric testing to detect or confirm hearing impairment in school-age children in rural South African schools. Their combined model showed a 100% specificity rate and a moderate sensitivity rate of around 65%, which they claim to be affected by randomization error and the susceptibility of children being easily distracted.

Note that hearing screenings, diagnostic examinations, and rehabilitation interventions in speech audiometry (speech-in-noise or word-in-noise) must always be implemented in a language that is native to the listeners [16,35]. Screening an individual's hearing with a language different than their native language often leads to poorer performance when compared to native listeners [22,35]. Hence, to our knowledge, there are currently no versatile Arabic versions of word-in-noise tests that can facilitate and sup-port the deployment of hearing screening programs that can be utilized in hard-to-reach environments.

Therefore, this study aims to develop an Arabic version of a word-in-noise hearing test that is cost-effective and has the potential to support its deployment in a low-resource setting and during community-based outreach programs. Furthermore, having such a hearing screening tool can reduce the burden on millions of people that have hidden hearing problems or do not have easy access to quality healthcare.

The current study presents the development of the proposed Arabic word-in-noise using an easy-to-use custom-made MATLAB (MathWorks, Natick, MA, USA) program. The main features that were considered during the design of the proposed Arabic word-in-noise screening tool are its ease of use with minimal training for generalist healthcare personnel, usability for Arabic-speaking communities, short testing time with reliable screening results, and the use of simple and common monosyllabic Arabic words that can be used to screen school-age children, young adults, and elderly persons. The words in the proposed Arabic hearing screening test in this study are implanted in various types of background noise to incorporate everyday life scenarios. In addition, the words were presented at different SNR levels to enhance the accuracy of identifying possible hearing impairments.

2. Materials and Methods

2.1. Participants

Thirty bilingual male participants with normal hearing volunteered in this study. All participants had acquired the Arabic language from birth and learned the English language during their formal education for at least five years. The screened participants were 18 to 33 years old, with a mean of 22.57 ± 2.42 SD years. All participants reported no history of neurological disorders, hearing loss, or hearing difficulties. Additionally, the screened participants confirmed that they had no history of cochlear and neural injuries or complaints of their cognitive functions. All participants had normal hearing as tested by pure tone audiometry through a smartphone application (uHear), which was validated for audiometry testing [29]. The uHear automated pure tone audiometry test was conducted via the same equipment across all participants (iPhone 11 and AirPods, Apple Inc., Cupertino, CA, USA). Participants were directed to follow the application instruction, and the tones were delivered at 500, 1000, 2000, 4000, and 6000 Octave Hz for the left and right ears. A consent form was required in the study for all the participants. The study was approved by the Institutional Review Board (I.R.B.) at the College of Applied Medical Sciences, King Saud University (K.S.U.) (I.R.B. Approval number: CAMS 029-3940).

2.2. Arabic Word Selection Process

To choose proper words to be used in this test, a total of 220 monosyllabic Arabic words were selected. The monosyllables were chosen based on their familiarity [11]. Seventy-five words of the total words (34%) were provided by the Department of Audiology at King Faisal Specialized Hospital and Research Center (KFSHRC), and the remaining 145 words were identified from school materials. The complete list of the Arabic words was then sent to two faculty members within K.S.U. to evaluate the words' complexity and familiarity with a four-point scale. The experts were asked to rate each word choice with "familiar and simple", "unfamiliar and complex", "familiar but complex", and "unfamiliar and complex". Next, words were ordered based on their familiarity and simplicity score. Eighty-four words that had the agreement of both evaluators for being familiar and simple were used as testing words in the current study (all of the words were in the KFHRC word list). Words that were evaluated by at least one of the evaluators as familiar and simple were used as choice words during the test after matching their homogeneity with testing words as far as possible. Finally, words that were rated as either unfamiliar and complex or familiar, but complex were excluded from the study.

2.3. Stimuli

The proposed test consists of monosyllabic Arabic words spoken by a male talker and recorded by audio software (Audacity Platform, Oak Park, MI, USA) via a commercially available microphone (Saramonic SR-ULM5, White Plains, NY, USA). Furthermore, 48 monosyllabic, phonetically balanced Arabic words were selected for test development. The Arabic words were randomly divided into six word lists (each with eight monosyllabic words). The words in the first three lists were embedded in white background noise (24 monosyllabic Arabic words), and the words in the remaining lists were embedded in multi-talker babble background noise (24 monosyllabic Arabic words).

Seven SNR conditions were presented for each word list in both noise types. The first word in each list was presented without background noise, and the remaining seven words were delivered with different SNR conditions. The SNR variations of the proposed Arabic word-in-noise screening test were +15, +10, +8, +6, +4, +2, and 0 dB (Figure 1). All the audio files, consisting of both the speech and background noise, have a duration of 8 s with a sampling frequency of 44,100 Hz.

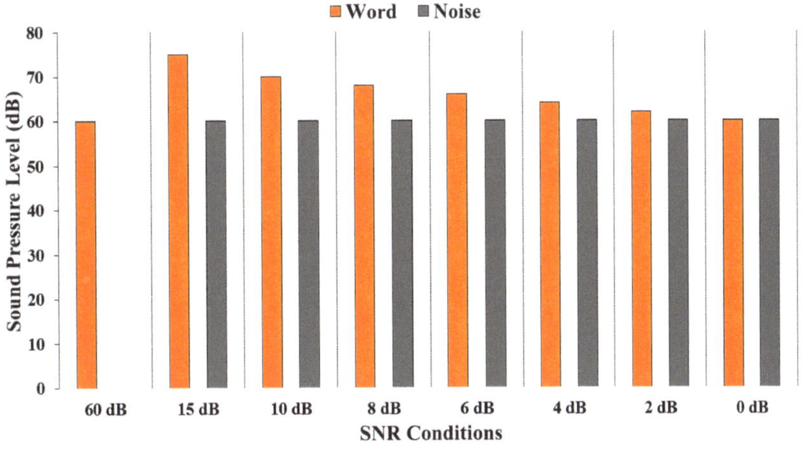

Figure 1. Illustration of the SNR conditions presented for both white noise and multitasker babble.

2.4. Experimental Procedure

Before starting each testing session, noise and speech level intensities were measured and calibrated using a digital sound level meter (Extech 407730, Waltham, MA, USA) to ensure that the ambient noise in the testing room (college research laboratory) was within an acceptable environment noise range (below 50 dB SPL) during testing, which is an important factor for such a mobile and outside of clinic hearing screening tool.

The testing procedure was administered with a headphone set (Sennheiser HD-280 pro, Wedemark, Germany), which was verified for audiometry testing [33,36]. Additionally, the output sound volume of the personal computer was measured and adjusted accordingly to ensure that the noise files used in this experiment were fixed at 60 dB using the digital sound meter before each session.

The flowchart of the proposed Arabic word-in-noise hearing assessment tool is shown in Figure 2. During each testing session, all participants were instructed to identify the presented Arabic words in both white noise lists (Word lists 1, 2, and 3) and multi-talker babble noise (Word lists 4, 5, and 6). Participants were instructed to listen carefully to the presented words. Moreover, the words in each list were delivered with a gradual decrease in the SNR for both the white and multi-talker babble noise (+15 to 0 dB), while the noise intensity was kept constant at 60 dB across all SNR conditions. Subjects were encouraged to guess the correct word from a multiple-choice list after each word. Participants were also able to select the last choice, "I do not know—لا أعلم" only if they did not perceive the word. In addition, the subjects would be excluded from the study if they failed to identify the first word in each word list, as they were presented without noise, and the session would be terminated. The order of the presented noise type in each session and the word lists was randomly assigned and counterbalanced across subjects to avoid an order effect.

The acquired responses were given a score of zero for wrong word identification and a score of one for correct word identification. In addition, an overall average score of three word lists was calculated for each noise type. The percent-correct word identification score was then computed and used for statistical analysis.

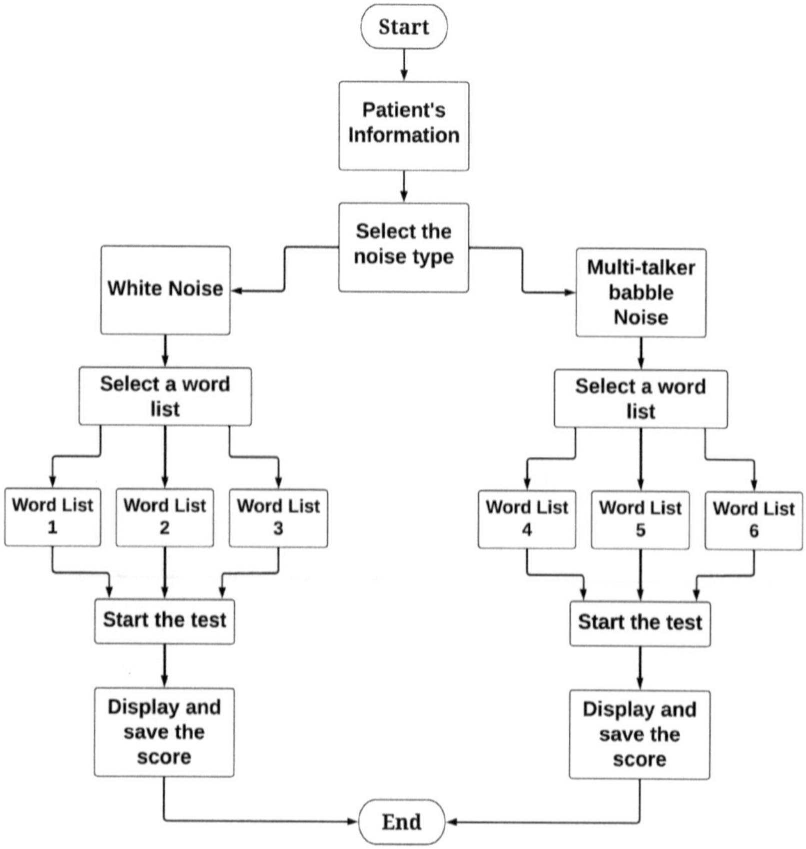

Figure 2. Flowchart of the proposed Arabic word-in-noise screening test.

3. Results

The pure tone audiometry results acquired by the "uHear" phone application confirmed that all subjects had normal hearing in this with normal hearing sensitivity in both ears at 0.5, 1, 2, 4, and 6 octave kHz.

All participants completed the proposed Arabic word-in-noise test with the two types of noise (white noise and four multi-talker babble noise). The average testing time for the Arabic word-in-noise test was seven minutes (SD 2.14).

The percent-correct scores ranged from 100% to 54.17% for the white background noise and 91.57% to 50% for the multi-talker babble background noise (Figure 3). Furthermore, the across-subject and across-list mean percent-correct word recognition scores were 80.14% (SD 9.83) and 74.17% (SD 10.69) for white and multi-talker babble background noise, respectively. A statistically significant difference was seen between the resulting scores from white noise and multi-talker babble noise (paired-samples t-test, $t\ (29) = 3.61$, $p = 0.001$) (Figure 4).

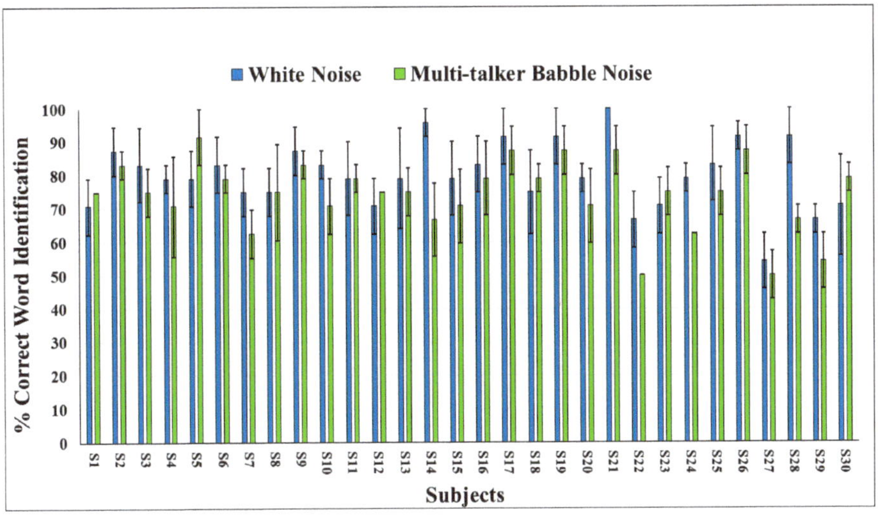

Figure 3. A cross-list mean (±SEM) percent-correct word identification score for each participant for both white noise and multi-talker babble background noise.

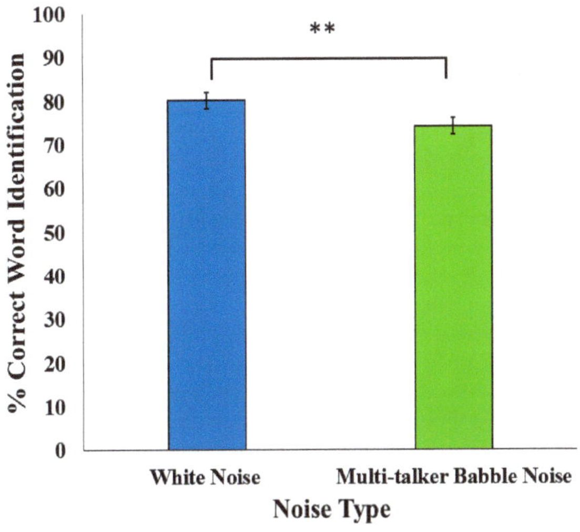

Figure 4. The across-subjects, across-list mean (±SEM) percent-correct word identification score for white noise and multi-talker babble background noise. (** $p < 0.01$).

4. Discussion

The primary goal of this research is to create and validate a computerized Arabic word-in-noise speech perception screening tool that can be used outside of the clinical setting. Outside-clinic tests are essential as they can be effective in rural areas and under political, security, or health pandemic constraints where proper healthcare services are hard to reach [33,37,38].

Developing and validating a speech perception screening tool depends on several factors, such as test duration, linguistics, and the type and intensity of noise. This study validates a proposed Arabic word-in-noise assessment test, reporting the advantages of a short testing time and affordable hardware. The proposed test also encompasses several SNR levels that vary in difficulty to enhance the accuracy of assessing speech intelligibility.

Additionally, there are two main categories of background noise, known as masking noises: stationary or fluctuant [39]. Noise with a changing intensity over time is fluctuant, whereas stationary noise has a fixed intensity over time. Furthermore, speech-in-noise audiometry tests often incorporate white noise or speech spectrum noise as a masking noise [39,40]. Although it does not reflect daily life situations, stationary white noise is most commonly used in speech-in-test clinical testing for its simplicity and ease of calibration [39]. In contrast, spectrum noise, such as multi-talker babble noise, is more representative of daily life situations. Thus, the two types of background noises were tested in this pilot study to investigate their impact on speech perception in healthy individuals. The results indicate that the performance of people with normal hearing was significantly higher in the presence of white noise (80.14%) compared to multi-talker babble noise (74.17%). This implies that listeners with normal hearing are affected differently by the type of background noise, which is in line with other studies showing that word recognition scores are lower in the presence of multi-talker babble noise [41–43].

The results of the present investigation also suggest that the use of white background noise is less challenging and more consistent compared to multi-talker babble noise, at least for Arabic-speaking adults with normal hearing. The use of white background noise could also be more favorable over multi-talker babble noise as the test is developed to be used out of the sound booth and clinical setup where environmental noise is less controlled. However, further investigations are needed to confirm the effect of background noise in diverse practical settings.

In comparison with the few existing Arabic speech-in-noise and word-in-noise tests (Arabic matrix test, Speech Perception of Words in Noise test, and the PAAST), the versatile Arabic word-in-noise test proposed in this study has some design advantages. First, the proposed test is a word-based screening tool that is much easier to complete and comprehend than the sentence-based tests used in the Arabic matrix test. Additionally, this versatile test has its testing words embedded in varying SNR levels, which enable primary care personnel to have a better screening result compared to the paradigm of a zero SNR variation between the presented word and the background noise as in the Speech Perception of Words in Noise test. Finally, the PAAST screening test has a limited targeted population where the word lists in the study can be recognized by school-age children, youth, and elderly populations.

The current study had some limitations, including the lack of enrollment of an age-matched group with hearing loss. Thus, further studies are required on age-matched individuals with hearing loss to validate the sensitivity and specificity of the versatile Arabic word-in-noise hearing screening tool used in this study. Additionally, the recruited sample size was relatively small and included only male participants. However, several studies have indicated a significant gender effect on speech intelligibility [44,45]. Moreover, the stimulation paradigm used in this study used a binaural approach in which words were delivered to both ears simultaneously. This technique might be a disadvantage in hearing screening if one ear has a hearing deficiency. That is, the binaural paradigm might mask a single-ear deficiency. Monaural stimulation for each ear separately could enhance the performance and detection of a unilateral hearing difficulty. Furthermore, although the performance of the uHear mobile application to assess hearing sensitivity has been validated in previous studies [29,46], performing a clinical pure-tone audiometry test and an otological examination prior to the screening with the Arabic WIN test would eliminate the possibility of existing of hearing issues.

5. Conclusions

In summary, the present results demonstrate the feasibility of developing a versatile Arabic word-in-noise screening tool, which has the potential to be used effectively to screen for speech perception difficulties. The tool can recommend that subjects scoring below a certain threshold consult a professional hearing specialist. Determination of the threshold level requires more investigation. In addition, further testing of the proposed hearing screening tool on different age groups is needed to identify and validate its efficacy and the best-targeted age group.

Author Contributions: Conceptualization, K.A. and A.O.A.; methodology, K.A., A.O.A. and A.F.A.; software, A.F.A.; validation, K.A., A.A., A.O.A. and A.F.A.; formal analysis, A.F.A. and A.O.A.; investigation, K.A., A.A. and A.O.A.; data curation, A.A. and A.O.A.; writing—original draft preparation, K.A., A.F.A. and A.O.A.; writing—review and editing, A.A., A.O.A. and K.A.; visualization, A.F.A. and A.O.A.; supervision, K.A. and A.A.; project administration, K.A. and A.A. All authors have read and agreed to the published version of the manuscript.

Funding: The authors extend their appreciation to the Deputyship for Research & Innovation, Ministry of Education in Saudi Arabia, for funding this research work through the project number IFKSURG-1171.

Institutional Review Board Statement: The study was conducted in accordance with the Declaration of Helsinki, and approved by the Institutional Review Board of the College of Applied Medical Sciences, King Saud University (CAMS 029-3940).

Informed Consent Statement: Informed consent was obtained from all subjects involved in the study.

Data Availability Statement: The data that support the findings of this study are available on request from the corresponding author [K.A.].

Conflicts of Interest: The authors declare no conflict of interest.

References

1. World Health Organization. Deafness and Hearing Loss. Available online: https://www.who.int/news-room/fact-sheets/detail/deafness-and-hearing-loss (accessed on 30 November 2022).
2. Tuczyńska, M.; Staszewski, R.; Matthews-Kozanecka, M.; Baum, E. Impact of Socioeconomic Status on the Perception of Accessibility to and Quality of Healthcare Services during the COVID-19 Pandemic among Poles—Pilot Study. *Int. J. Environ. Res. Public Health* **2022**, *19*, 5734. [CrossRef]
3. Knudsen, L.V.; Öberg, M.; Nielsen, C.; Naylor, G.; Kramer, S.E. Factors Influencing Help Seeking, Hearing Aid Uptake, Hearing Aid Use and Satisfaction with Hearing Aids: A Review of the Literature. *Trends Amplif.* **2010**, *14*, 127–154. [CrossRef] [PubMed]
4. Garramiola-Bilbao, I.; Rodríguez-Álvarez, A. Linking hearing impairment, employment and education. *Public* **2016**, *141*, 130–135. [CrossRef] [PubMed]
5. Jung, D.; Bhattacharyya, N. Association of Hearing Loss with Decreased Employment and Income among Adults in the United States. *Ann. Otol. Rhinol. Laryngol.* **2012**, *121*, 771–775. [CrossRef]
6. Li, C.-M.; Zhang, X.; Hoffman, H.; Cotch, M.F.; Themann, C.L.; Wilson, M.R. Hearing Impairment Associated with Depression in US Adults, National Health and Nutrition Examination Survey 2005–2010. *JAMA Otolaryngol. Neck Surg.* **2014**, *140*, 293. [CrossRef]
7. Davis, A.; Smith, P.; Ferguson, M.; Stephens, D.; Gianopoulos, I. Acceptability, benefit and costs of early screening for hearing disability: A study of potential screening tests and models. *Health Technol. Assess.* **2007**, *11*, 1–294. [CrossRef]
8. Hatton, J.L.; Rowlandson, J.; Beers, A.; Small, S. Telehealth-enabled auditory brainstem response testing for infants living in rural communities: The British Columbia Early Hearing Program experience. *Int. J. Audiol.* **2019**, *58*, 381–392. [CrossRef] [PubMed]
9. Alhussaini, K.; Bohorquez, J.; Delgado, R.E.; Ozdamar, O. Auditory brainstem, middle and late latency responses to short gaps in noise at different presentation rates. *Int. J. Audiol.* **2018**, *57*, 399–406. [CrossRef]
10. Behar, A. Audiometric Tests without Booths. *Int. J. Environ. Res. Public Health* **2021**, *18*, 3073. [CrossRef]
11. Ashoor, A.A.; Prochazka, T. Saudi Arabic Speech Audiometry. *Int. J. Audiol.* **1982**, *21*, 493–508. [CrossRef]
12. Tatham, M.; Morton, K. *A Guide to Speech Production and Perception*; Edinburgh University Press: Edinburgh, UK, 2011. [CrossRef]
13. Chermak, G.D.; Hall, J.W.; Musiek, F.E. Differential Diagnosis and Management of Central Auditory Processing Disorder and Attention Deficit Hyperactivity Disorder. *J. Am. Acad. Audiol.* **1999**, *10*, 289–303. [CrossRef] [PubMed]
14. Inoue, T.; Higashibara, F.; Okazaki, S.; Maekawa, H. Speech perception in noise deficits in Japanese children with reading difficulties: Effects of presentation rate. *Res. Dev. Disabil.* **2011**, *32*, 2748–2757. [CrossRef] [PubMed]
15. Haesen, B.; Boets, B.; Wagemans, J. A review of behavioural and electrophysiological studies on auditory processing and speech perception in autism spectrum disorders. *Res. Autism Spectr. Disord.* **2011**, *5*, 701–714. [CrossRef]
16. Portnuff, C.; Bell, B. Effective Use of Speech-in-Noise Testing in the Clinic. *Hear. J.* **2019**, *72*, 40. [CrossRef]
17. Pisoni, D.B. Word Identification in Noise. *Lang. Cogn. Process.* **1996**, *11*, 681–688. [CrossRef]
18. Holder, J.; Sheffield, S.W.; Gifford, R. Speech Understanding in Children with Normal Hearing. *Otol. Neurotol.* **2016**, *37*, e50–e55. [CrossRef]

19. Schafer, E.C. Speech perception in noise measures for children: A critical review and case studies. *J. Educ. Audiol.* **2010**, *16*, 4–15.
20. Mendel, L.L.; Widner, H. Speech perception in noise for bilingual listeners with normal hearing. *Int. J. Audiol.* **2015**, *55*, 126–134. [CrossRef]
21. Hjertman, H.; Hällgren, M.; Mäki-Torkko, E.; Stenfelt, S. The Swedish hearing in noise test for children, HINT-C. *Int. J. Pediatr. Otorhinolaryngol.* **2020**, *141*, 110509. [CrossRef]
22. Wilson, R.H. Development of a Speech-in-Multitalker-Babble Paradigm to Assess Word-Recognition Performance. *J. Am. Acad. Audiol.* **2003**, *14*, 453–470. [CrossRef]
23. Kollmeier, B. The Arabic Matrix Test: Cochlear and HörTech Introduce Reliable Speech Audiometry to the Arabic Speaking World. 2014. Available online: http://www.cochlear.com/wps/wcm/connect/uk/for-professionals/sound-connection/the-arabic-matrix-test (accessed on 22 August 2019).
24. Rahman, T.T.A. Development and standardization of Arabic words in noise test in Egyptian children. *Int. J. Pediatr. Otorhinolaryngol.* **2018**, *108*, 1–7. [CrossRef] [PubMed]
25. Alkahtani, R.A. The Development of the Paediatric Arabic Auditory Speech Test for iPad. *ENT & Audiology News*. 1 May 2019. Available online: https://www.entandaudiologynews.com/reviews/tech-reviews/post/the-development-of-the-paediatric-arabic-auditory-speech-test-for-ipad (accessed on 30 November 2022).
26. Irace, A.L.; Sharma, R.K.; Reed, N.S.; Golub, J.S. Smartphone-Based Applications to Detect Hearing Loss: A Review of Current Technology. *J. Am. Geriatr. Soc.* **2020**, *69*, 307–316. [CrossRef] [PubMed]
27. Corona, A.P.; Ferrite, S.; Bright, T.; Polack, S. Validity of hearing screening using hearTest smartphone-based audiometry: Performance evaluation of different response modes. *Int. J. Audiol.* **2020**, *59*, 666–673. [CrossRef] [PubMed]
28. Yousuf, S.; Hussein, S.Y.; Swanepoel, D.W.; Mahomed, F.; de Jager, L.B. Community-based hearing screening for young children using an mHealth service-delivery model. *Glob. Health Action* **2018**, *11*, 1467077. [CrossRef] [PubMed]
29. Anuar, K.B.; Rani, M.; Hitam, S.B.; Noh, A.B. Exploring Uhear Hearing Application as a mobile screening tool for the underprivileged. *Rawal Med. J.* **2018**, *43*, 717–720.
30. Davis, L.; Hwa, T.P.; Eliades, S.J.; Bigelow, D.C.; Ruckenstein, M.J.; Brant, J.A. Pure tone audiometry as assessed by a commercially-available mobile phone application compared to formal audiometry. *Am. J. Otolaryngol.* **2022**, *43*, 103552. [CrossRef]
31. Rye, P.; Andersen, R.O.; Ravn, G.; Hammershøi, D. Comparison of Speech Reception Thresholds for diotic, dichotic and antiphasic headphone presentations of digits-in-noise triplets using Dantale I material. In *Conference Proceedings of the Euroregio/BNAM 2022 Joint Acoustic Conference*; The European Acoustics Association (EAA): Lyon, France, 2022; pp. 383–391.
32. Bauer, M.A.; Sales, A.; Teixeira, A.R.; Morsch, P.; Lessa, A.H.; Bós, J.G. Development and accuracy of a hearing screening application. *Braz. J. Otorhinolaryngol.* **2021**, *87*, 711–717. [CrossRef]
33. Alhussaini, K.; Saleh, S.; Aleid, A.; Alkhalaf, S.; Badghaish, R.; Altinawi, A.; Alwasel, A. Software Application toward Accessible Hearing Care Assessment: Gap in Noise Test. *J. Health Eng.* **2022**, *2022*, 1–6. [CrossRef]
34. Govender, S.M.; Mars, M. Assessing the efficacy of asynchronous telehealth-based hearing screening and diagnostic services using automated audiometry in a rural South African school. *South Afr. J. Commun. Disord.* **2018**, *65*, 9. [CrossRef]
35. Zokoll, M.A.; Hochmuth, S.; Warzybok, A.; Wagener, K.C.; Buschermöhle, M.; Kollmeier, B. Speech-in-Noise Tests for Multilingual Hearing Screening and Diagnostics1. *Am. J. Audiol.* **2013**, *22*, 175–178. [CrossRef]
36. Van Der Aerschot, M.; Swanepoel, D.W.; Mahomed-Asmail, F.; Myburgh, H.C.; Eikelboom, R. Affordable headphones for accessible screening audiometry: An evaluation of the Sennheiser HD202 II supra-aural headphone. *Int. J. Audiol.* **2016**, *55*, 616–622. [CrossRef]
37. Alhussaini, K.; Alkhalaf, S.; Aleid, A. Software application for auditory temporal resolution screening outside the clinic. In Proceedings of the 2020 IEEE 5th Middle East and Africa Conference on Biomedical Engineering (MECBME), Amman, Jordan, 27–29 October 2020; IEEE: Piscataway, NJ, USA, 2020. [CrossRef]
38. Ceccato, J.-C.; Duran, M.-J.; Swanepoel, D.W.; Smits, C.; De Sousa, K.C.; Gledhill, L.; Venail, F.; Puel, J.-L. French Version of the Antiphasic Digits-in-Noise Test for Smartphone Hearing Screening. *Front. Public Health* **2021**, *9*, 725080. [CrossRef]
39. Reynard, P.; Lagacé, J.; Joly, C.-A.; Dodelé, L.; Veuillet, E.; Thai-Van, H. Speech-in-Noise Audiometry in Adults: A Review of the Available Tests for French Speakers. *Audiol. Neurotol.* **2021**, *27*, 185–199. [CrossRef]
40. Wagener, K.C.; Brand, T. Sentence intelligibility in noise for listeners with normal hearing and hearing impairment: Influence of measurement procedure and masking parameters La inteligibilidad de frases en silencio para sujetos con audición normal y con hipoacusia: La influencia del procedimiento de medición y de los parámetros de enmascaramiento. *Int. J. Audiol.* **2005**, *44*, 144–156. [CrossRef]
41. Lee, S.H.; Shim, H.J.; Yoon, S.W.; Lee, K.W. Effects of Various Background Noises on Speech Intelligibility of Normal Hearing Subjects. *Korean J. Otolaryngol. Neck Surg.* **2009**, *52*, 307. [CrossRef]
42. Goldsworthy, R.L.; Markle, K.L. Pediatric Hearing Loss and Speech Recognition in Quiet and in Different Types of Background Noise. *J. Speech, Lang. Hear. Res.* **2019**, *62*, 758–767. [CrossRef]
43. Kattel, B.P.; Fasanya, B.K.; Letowski, T.R.; Hargrove, S.K. The effect of types of background noise on the acceptable noise levels of individuals with normal hearing. In Proceedings of the 13th Annual International Conference of Industrial Engineering. Theory, Applications and Practice, Las Vegas, NV, USA, 7–10 September 2008.
44. Yumba, W.K. Influences of listener gender and working memory capacity on speech recognition in noise for hearing aid users. *Speech Lang. Hear.* **2020**, *25*, 112–124. [CrossRef]